# Metallography as a Quality Control Tool

# Metallography as a Quality Control Tool

Edited by
## James L.McCall
*Battelle-Columbus Laboratories*
*Columbus, Ohio*

## and
## P.M.French
*Westinghouse Electric Corporation*
*Madison, Pennsylvania*

Plenum Press · New York and London

Library of Congress Cataloging in Publication Data

Symposium on Metallography as a Quality Control Tool, Tamiment, Pa., 1979.
 Metallography as a quality control tool.

 Proceedings of the symposium held July 8—9, 1979.
 Includes indexes.
 1. Metallography—Congresses. 2. Quality control—Congresses. I. McCall, James L.
II. French, Peter Michael, 1935-    III. Title.
TN689.2.S878 1979                   669'.95                              80-388
ISBN 0-306-40423-0

Proceedings of a symposium on Metallography as a Quality Control Tool,
sponsored by the American Society for Metals and the International
Metallographic Society, held in Tamiment, Pennsylvania, July 8—9, 1979


© 1980 Plenum Press, New York
A Division of Plenum Publishing Corporation
227 West 17th Street, New York, N.Y. 10011

Printed in the United States of America

## PREFACE

Quality control has been described as a system for verifying and maintaining a desired level of quality in a product or process by careful planning, continued inspection, and corrective action where required. With many of today's products, there is an ever increasing demand for improved reliability during service. This in turn necessitates the use of a wide range of control techniques — some very sophisticated and complex — not only to verify the quality of the final product but also to monitor that the fabrication processes are under control. Furthermore, with certain industries, quality control of the final product is of paramount importance because of the needs for its reliable and safe operation under arduous and sometimes dangerous conditions. Metallography often serves as an invaluable quality control tool and can provide information not normally attainable by more conventional procedures. It often supplements both destructive techniques, e.g., mechanical testing, as well as non-destructive procedures, e.g., as radiography, ultrasonic testing, and dye-penetrant inspection. Furthermore, metallographic inspection utilizes a wide range of techniques ranging from conventional optical microscopy to more sophisticated procedures such as scanning electron microscopy, X-ray spectroscopy, and Auger electron spectroscopy. In some industries, metallography also is employed during maintenance, field inspection, and overhaul of components. It is often a primary tool in the analysis of a failed component by helping pinpoint the cause and in preventing further failures by uncovering shortcomings in the processing of the materials involved, through characterization of defects, and by revealing problems introduced during manufacture.

In recognition of the important role that metallography plays in the broad area of quality control, the absence of a text that specifically discusses this subject, and the belief that communication of information on the subject would be of technical interest, The International Metallographic Society and The American Society for Metals co-sponsored a Symposium on the subject. The intent was to bring together world-recognized authorities working in various aspects of metallography and quality control to share methods they have developed and used, to review the importance in their particular area of application, to discuss the data obtained and finally recognize conclusions that can be drawn. The symposium, entitled "Metallography as a Quality Control Tool", was held at Tamiment, Pennsylvania, U.S.A., July 8–9, 1979. It followed four earlier symposia co-sponsored by the same two societies on other subjects of interest to the metallographic community, namely; Microstructural Analysis—Tools and Techniques (1972), Metallographic Specimen Preparation—Optical and Electron Microscopy (1973), Interpretative Techniques for Microstructural Analysis (1975), and Metallography in Failure Analysis (1977).

v

The wide-spread interest in the symposium, specifically shown by the large attendance from around the world and enthusiastic participation, has encouraged us to publish all of the formally presented papers. These papers comprise the current volume. Our hope is that these proceedings will serve as a useful reference for individuals active either full, or part-time, in this field.

Organizing a symposium of this size and type would not have been possible without the combined efforts of many individuals. To all we owe a deep debt of gratitude, but, especially we want to mention G.F. Vander Voort, General Chairman of the 1979 International Metallographic Society Convention, of which this symposium was a part. The cooperation of both co-sponsoring societies was assured by the efforts of several individuals, most directly Dr. E.J. Meyers and Mr. J. Devis of the American Society for Metals and Mr. R.J. Gray of the International Metallographic Society. We also give a special thanks to Connie McCall for putting the entire proceedings in a uniform format and typing them in camera-ready form. Finally, we thank all the authors and session chairmen without whose enthusiastic participation the symposium obviously would not have been possible.

James L. McCall
Battelle — Columbus Laboratories

P.M. French
Westinghouse Electric Corporation

## CONTENTS

# INCLUSION MEASUREMENT

George F. Vander Voort*

## INTRODUCTION

Nonmetallic inclusions are present in all metals, irrespective of the type or processing procedure. Since inclusions significantly influence properties and behavior of materials, they have been studied extensively. Primarily because of the much greater use of steel in critical applications, most of such research has centered on inclusions present in steel rather than in other materials. Regardless of the material involved, quantification of inclusion content is important in many quality control or research studies. This paper deals with the myriad of methods proposed over the years to assess inclusion content.

Inclusions can be catalogued in a number of ways. Inclusion origin, for example, produces two categories: exogenous and endogenous, i.e., indigenous. Exogenous inclusions come from external sources such as refractories. Indigenous inclusions arise from internal sources; they are, for example, products of deoxidation or precipitation of sulfides. Since an inclusion by definition refers to any foreign body enclosed within the mass of an object, the word *inclusion* should be used only for those from exogenous sources. Sulfides or deoxidation products could be considered nonmetallic phases present in a material. However, the use of the term inclusion to encompass particles from indigenous and exogenous sources has been widely accepted.

Indigenous inclusions can be subdivided as oxides and sulfides. Nitrides and carbides that are visible in the optical microscope are not dealt with as inclusions, because they usually have characteristics more akin to those of metals rather than of nonmetallics.

Inclusions may also be classified according to size, i.e., macroscopic or microscopic. There is no clear dividing line between these two size ranges. Most exogenous inclusions are generally large in size and are distributed haphazardly. Indigenous inclusions are usually small, generally less than 100 $\mu$m in diameter (in the as-cast condition) and, according to one study[1] may be as small as 30-40 nm. Exogenous inclusions as small as 10 $\mu$m are observed. Hence, the size range of indigenous and exogenous inclusions overlap and one cannot simply state that all indigenous inclusions are microscopic and all exogenous inclusions are macroscopic.

The literature use of certain terms in place of the words *nonmetallic inclusion* creates misconceptions. Inclusions are sometimes referred to as *dirt*, and, in fact, one chart for rating inclusions[2] is referred to as the "Dirt Chart". While it is

*Bethlehem Steel Corporation, Research Department, Bethlehem, Pennsylvania, USA.

1

obviously possible to entrap some debris in ingots, for example, because of inadequate mold cleaning prior to teeming, the amount of such material is small compared to the quantity of inclusions of indigenous origin. The "Dirt Chart" actually shows two types of indigenous inclusions:  malleable (sulfides and many silicates) and brittle (aluminates).

Inclusions have often been widely referred to as *slag particles*, regardless of their origin.  Some slag-related material may be present in steels or some of the inclusions in steel may have reacted with slag;  however, only a small portion of the total inclusion population are really slag particles as such.  The one exception would be wrought iron, where several percent of a specially prepared slag are mechanically mixed into relatively  pure molten iron.  The "Dirt Chart" is also referred to as the "Slag/Oxide Chart", where the malleable inclusions are referred to as slag particles and the brittle inclusions are the oxides.  Clearly, such usage creates considerable confusion.  Another example of unrigorous usage is the chart developed by Diergarten[3,4] where the rating categories are *sulfide slags, brittle slags* and *oxide slags.*

The terms steel *cleanliness* or *microcleanliness* also present a problem.  A clean steel is one whose surface is free of dirt, grease, etc.  This term stems from the reference to inclusions as dirt;  hence, a clean steel is one free of *dirt.*  Some publications use the term *purity*, but this term refers not to inclusions but to the presence of elements other than the primary element or compound.

The use of ambiguous terminology poses a problem for the automatic retrieval of documents by means of key words.  Ambiguous key words will thus produce many references that are not germane to the subject of inclusions.

The distribution of inclusions varies according to their origin and type.  Exogenous inclusions are not uniformly distributed throughout the cast or wrought product.  Hence, the chance of detecting them on a randomly chosen plane of polish is extremely small.  Macroscopic inspection procedures must be used to locate these particles before they can be studied microscopically.  Indigenous inclusions, however, tend to be distributed in a more uniform manner, although their concentration does vary with location.  In general, sulfides tend to segregate towards the centerline and top of an ingot, whereas oxides tend to segregate towards the centerline and bottom of the ingot.  Subsurface oxides, often due to secondary oxidation effects, can frequently be observed.  All these factors must be considered in planning sampling procedures for quality control studies.

## METHODS FOR DETECTING INCLUSIONS

Due to variation in size and distribution of inclusions, a number of procedures or methods have been developed to detect inclusions and assess their concentration:
Macroscopic Methods
- Macroetch or hot acid etch test
- Contact printing
- Fracture test
- Magnetic particle inspection
- Ultrasonics

Microscopic Methods
- Optical microscopy
- Microradiography
- Transmission or scanning electron microscopy

Chemical Methods
- Isolation of residues
- Analytical methods for oxygen and sulfur.

## MACROSCOPIC METHODS

The macroscopic methods are particularly useful, since they enable a large area or volume to be studied in a short time. However, they are generally not suitable for determining the type of inclusion —— of considerable importance because different inclusions have different effects on properties. Hence, macroscopic methods must be followed by microscopic tests.

### Hot Acid Etch Test

Macroetching using relatively strong acid solutions has been widely used to evaluate steel quality. Generally, a disk about one-half inch thick is cut from a billet or bloom and etched using a 1 to 1 solution of hydrochloric acid in water for 15—45 minutes at about 160 F. The sample is rinsed and dried before examination (visual to about 10X). The results are expressed in qualitative terms only and may be compared to a series of standard pictures[5—9]. Some of these standards include examples of disks containing indications due to inclusions. Figure 1 shows an etch disk revealing inclusions concentrated in an area near the surface, a condition referred to as "subsurface dirt". Microscopic examination of this area revealed manganese oxide and manganese silicate. These inclusions contained considerable chromium from the steel and a minor amount of aluminum. Another example is given in Fig. 2. This is an etch disk removed from the top of a billet that contained entrapped slag as a result of an insufficient top discard. Microscopic examination and microprobe analysis confirmed that the large, exogenous inclusions were entrapped slag.

### Contact Printing

Contact printing techniques have been developed to reveal the distribution of sulfur, phosphorus, oxide and lead. Of these methods, only the sulfur, or Baumann, print is widely used. To make a sulfur print, a piece of bromide-type photographic paper is immersed in an aqueous 2% acid solution and then carefully laid on the surface of a smooth-ground steel disk for 1—2 minutes. The paper is removed, washe in water, fixed in hypo, washed and dried. The distribution of sulfur and the presence of large sulfides or clusters of sulfides is clearly shown. Under standardized conditions, an experienced operator can make an approximate estimate of the sulfur content. An example of a sulfur print of a carbon steel pinion is given in Fig. 3. The steel contained 0.031% sulfur. The sulfur print is relatively dark and numerous large black (actually brown) spots can be seen that are indicative of large sulfides or clusters. Figure 4 shows the same disk after hot acid etching. Some of the larger sulfides can be observed.

### Fracture Test

For certain steel products the mill metallurgist will harden the disk used for the hot acid etch test (steel must be capable of being hardened to about 60 HRC). The disk is then fractured and blued by oxidizing the fracture on a hot plate or in a tempering furnace, 500—700 F usually being adequate. Inclusion stringers are readily observed (Fig. 5) on the fracture face (white streaks against a blue background). The fracture plane should be longitudinal with respect to the hot-working axis.

The fracture test is described in References 10 to 12. The ISO standard provides guidelines for both qualitative and quantitative inclusion measurement on blued fractures. The qualitative examination is conducted by comparing the sample with

Fig. 1. Hot acid etching of this disk re-vealed inclusions concentrated in an area below the surface (arrow). Some of the large pits in the central region may be due to inclusions.

Fig. 2. Hot acid etching of this disk taken from a billet location near the top of the ingot revealed entrapped slag (light colored substance).

Fig. 3. Sulfur print of a disk cut from a gear. The steel contained 0.031% sul-fur.

Fig. 4. Photograph of the disk shown in Fig. 3 after hot acid etching. Note the similarity of the details revealed by the two techniques.

Fig. 5.  Photograph of fractured and blued etch disk revealing inclusion stringers.

ten schematic diagrams (Fig. 6)  and noting the distribution of the stringers (core, surface or uniform distribution).  The quantitative examination requires counting of the inclusions and measuring the length and/or thickness.  The distribution of the inclusions according to the measured parameters is established according to the scheme given in Table 1.

Magnetic Particle Inspection

Ferromagnetic materials can be examined by magnetic particle inspection[10-12]. The machined surface of a test piece is placed in a magnetic field and coated with a liquid containing a suspension of ferromagnetic particles.  The presence of nonmetallic inclusions disrupts the magnetic field that causes attraction of the ferromagnetic particles and thus renders the inclusion visible.  Other imperfections (cracks, seams, laps, etc.) can also attract the powder.  These include cracks and other types of unconsolidated metal that are revealed by dye-penetrant inspection.

Examination surfaces can include the surface of the product after surface grinding, an axial core bar, a step-down test piece, or cylindrical test pieces machined from the product.  Specific details about specimen preparation can be found in References 10—12.  Indications representing inclusions can be recorded by photography or contact printing or can be sketched.

## TABLE 1

### Quantitative inclusion assessment — blue fracture method

ISO 3763

Inclusion Distribution Based on Length

| Symbol | Length, $\ell$ (mm) |
|--------|---------------------|
| $L_0$ | no macroscopic inclusions |
| $L_1$ | $1.0 \leq \ell \leq 2.5$ |
| $L_2$ | $2.5 < \ell \leq 5.0$ |
| $L_3$ | $5.0 < \ell \leq 10.0$ |
| $L_4$ | $\ell > 10.0$ |

(The number of inclusions less than 1 mm in length should be noted)

Inclusion Distribution Based on Thickness

| Symbol | Thickness, e (mm) |
|--------|-------------------|
| $T_0$ | no macroscopic inclusion |
| $T_1$ | $0.1 \leq e \leq 0.25$ |
| $T_2$ | $0.25 < e \leq 0.50$ |
| $T_3$ | $0.50 < e \leq 1.00$ |
| $T_4$ | $e > 1.00$ |

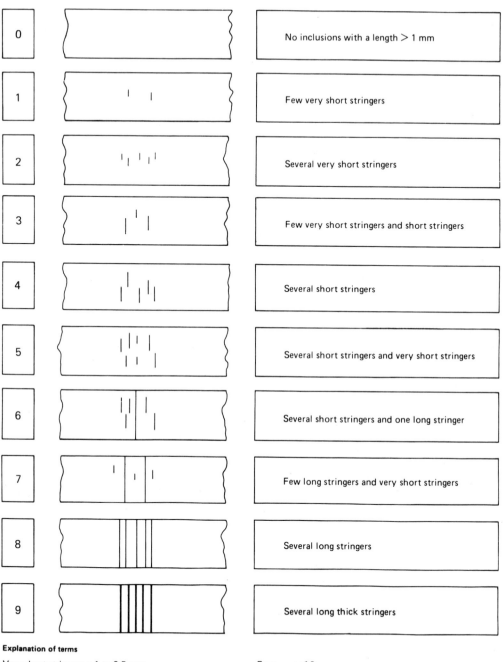

| | | |
|---|---|---|
| 0 | | No inclusions with a length > 1 mm |
| 1 | | Few very short stringers |
| 2 | | Several very short stringers |
| 3 | | Few very short stringers and short stringers |
| 4 | | Several short stringers |
| 5 | | Several short stringers and very short stringers |
| 6 | | Several short stringers and one long stringer |
| 7 | | Few long stringers and very short stringers |
| 8 | | Several long stringers |
| 9 | | Several long thick stringers |

**Explanation of terms**

| Very short stringers | : 1 to 2,5 mm | Few | : ⩽ 3 |
|---|---|---|---|
| Short stringers | : > 2,5 mm, ⩽ 5 mm | Several | : > 3 |
| Long stringers | : > 5 mm | Thick | : > 0,5 mm |

Fig. 6. Schematic diagrams in ISO 3763 for qualitative assessment of blue fracture test samples.

Test results are expressed in terms of frequency and severity. Frequency is the total number of indications in a given area (such as 40 sq. in.) or the number per square inch of surface examined (AMS 2301). Severity is the weighted value of the magnetic particle indications according to the scheme in Table 2. To obtain the severity number, multiply the number of indications in a given length class by the weight factor and sum up the results. The severity factor is expressed as the weighted value for a given area, usually 40 sq. in. but can be expressed as a value per square inch. For step-down test samples the severity is listed for each section diameter.

TABLE 2

Calculation of severity rating — magnetic particle indications — ASTM E45.

| Length of Inclusions | | Weight Factor |
|---|---|---|
| (inch) | (mm) | |
| Over 1/16 to 1/8 | 1.59 to 3.18 | 0.5 |
| Over 1/8 to 1/4 | 3.18 to 6.35 | 1 |
| Over 1/4 to 1/2 | 6.35 to 12.7 | 2 |
| Over 1/2 to 3/4 | 12.7 to 19.1 | 4 |
| Over 3/4 to 1 | 19.1 to 25.4 | 8 |
| Over 1 | >25.4 | 16 |

Ultrasonics

Ultrasonic inspection[13–20] is an exceptionally useful evaluation method, because it is nondestructive and can survey a large volume of material in a relatively short time. Hence, it is widely applied in quality control programs. Of course, the method is also useful for detecting features other than inclusions. Routine ultrasonic inspection techniques are adequate to detect the larger, more harmful inclusions, and laboratory equipment can map out the location of inclusions down to very small sizes. These techniques have the advantage of being able to cover a wide range of inclusion sizes and are particularly suitable for locating large exogenous inclusions, which can then be examined by other methods.

Ultrasonic waves are high-frequency mechanical vibrations introduced into the steel by means of a transducer. The waves travel through the steel until they encounter a discontinuity that creates a reflection. Both contact and immersion methods are utilized in the study of inclusions. The former method is used for routine quality control, whereas the latter method is performed in the laboratory. Ultrasonics can provide information on the size, location and frequency of indications. There is some ambiguity in the method, simply because one is not always certain that the indication studied was caused by an inclusion. Other material factors can introduce similar reflections.

## MICROSCOPIC METHODS

The major emphasis of this paper is on microscopic methods, because these methods enable the investigator to arrive at a more complete determination of the type, size, shape and distribution of inclusions than is possible with macroscopic methods.

Optical Microscopy

The optical microscope has been widely used[21—48] to examine polished met-
allographic samples for inclusion content.  The simplest method is to survey the
plane of polish and describe the types of inclusions present.  The inclusion content
can be rated according to a number of methods:
- Comparison with standard charts
- Counting methods
- Volume fraction measurement by lineal analysis or point counting
- Utilization of automatic or semiautomatic devices

These methods will be described in later sections.

Successful optical examination depends on the sampling procedures and the
quality of specimen preparation.  In an as-cast metal all sample orientations are basically equivalent.  However, radial orientations, i.e., parallel to the solidification direction, are preferred.  In a wrought specimen the polished surface is chosen parallel to
the hot-working axis, because the malleability of the inclusions is an important factor in identification and in quantification schemes.  For best results samples should
be hardened before polishing.  For routine quality control the mill metallurgist choo-
ses samples from the top and bottom of the first, middle and last ingots after they
are rolled to billets about 4 in. sq.  In some cases samples from the middle of the
ingot are also examined.  The specimens are taken from the mid-thickness location
and should have a surface area of about 1 sq. in.  Specific details regarding sampling
of steel products can be found in Reference 10.

Polishing of specimens for inclusion examination is critical, because inclusions
must be fully retained without smearing or "comet-tailing" and with a minimum of
scratches, pitting and staining.  If the volume fraction is to be determined, relief
must be minimized to avoid having it increase the apparent size of an inclusion.
Automatic polishing devices are most helpful in obtaining properly polished specimens.
Alternate polish and etching can be helpful in reducing surface deformation and metal smearing, especially for samples that cannot be hardened appreciably by heat
treatment.  A light picral or nital etch after polishing may also prove useful in shar-
pening the inclusion-metal interface.

A few studies[32—33] have utilized thin sections examined with transmitted
light.  The samples must be about 10 $\mu$m thick with both surfaces parallel.  Since
such sample preparation is tedious, this method is rarely employed.  Other infrequent-
ly used methods include the physical removal of the larger inclusions with a pointed
instrument, an *ultrasonic jackhammer* or an ultrasonic drill[49—51].  The extracted
particles are placed on a slide and examined with a petrographic microscope.

The majority of inclusion studies are conducted with bright field illumination,
which may be augmented with polarized light examination[44—48].  Most of the
known inclusion types have been examined with both bright-field and polarized light
(reflected or transmitted) and their color, reflectivity and response to polarized light
noted[23,30].  These techniques were widely used to identify inclusions prior to the
development of the electron microprobe and the scanning electron microscope (with
energy-dispersive analysis capability).

One of the oldest methods for identifying inclusions involves the use of various
etching reagents that selectively attack inclusion phases[26].  However, such procedures are tedious and are therefore rarely used now.

Ernest and Koenig[29] developed a technique for identifying inclusions through measurement of the optical properties of transparent inclusions in wrought steels. The specimen is sectioned at an angle of 30 degrees from the longitudinal direction. The surface is deep-etched so that the elongated inclusions protrude above the etched surface. The inclusions are examined using liquids of known indices of refraction to determine the refractive index of the inclusion.

Optical microscopy is invaluable in determining the size, shape and distribution of inclusions and in assessing the plasticity of inclusions. The size and distribution of indigenous inclusions are primarily a function of the freezing rate. However, some trends with respect to composition are observed. Since they do not have much chance for growth, inclusions that precipitate from the solid metal are small and randomly distributed. Inclusions that precipitate from the melt before solidification will coalesce and form larger particles.

Observations can also reveal the solidification or precipitation sequence and indicate when the inclusion formed with reference to the solidification process. Inclusions that precipitate from the melt as immiscible liquid globules are spherical, whereas inclusions that precipitate as solid particles just before or during solidification generally are angular or dendritic. The shape and location of various phases in a complex inclusion can indicate the sequence of precipitation.

Microradiography

Microradiography has been used in some inclusion studies[52–54]. The fact that a thin section (0.002 in.) of material must be employed creates some sample-preparation difficulty. The specimen is placed in contact with a suitable high-resolution photographic film, and the opposite side is exposed to monochromatic x-rays. The photographic image can be examined in a microscope with magnifications to about 200X. The choice of the radiation wavelength is a function of the mass-absorption coefficients of the elements present. The technique is ideal for studying leaded steels, because the mass-absorption coefficient for lead is much greater than iron for nearly all x-ray wavelengths. Hence, lead shows up as a bright white spot on the film. To detect manganese in sulfide inclusions, one uses cobalt $K_\alpha$ radiation followed by chromium $K_\alpha$ radiation. Manganese-rich inclusions in iron appear lighter than the iron matrix on the negative with cobalt radiation and darker than the matrix with chromium radiation. Holes appear darker than the matrix with either type of radiation.

## CHEMICAL METHODS

Although chemical methods are of course suited mainly for determining composition, at least one technique has been developed to provide size information as well.

Isolation of Residues

Inclusion type and content can be determined using a variety of chemical extraction methods[55–76]. While these date back to the work of Arnold and Read (1894), such methods tend to be used only sporadically in research and are not applied routinely in quality control programs. Complete separation of an uncontaminated residue is difficult if not impossible to achieve.

The chemical-extraction approach is most easily applied to thermodynamically stable oxides, such as occur in well-deoxidized steels. Various types of chemical and electrolytic methods have been developed. In general, a metallographic specimen or a sample of similar bulk is used. A small portion, typically about 1–20 grams, is dissolved from the specimen. The residue is removed from the attacking solution by washing, filtering and centrifuging. The residue can be analyzed by means of a variety of chemical analytical procedures and equipment. The percentage of oxides can be expressed as a function of the weight of the oxides in the residue to weight loss of the specimen during the dissolution process. Extraction techniques for sulfides are more difficult, because they are not as stable or as resistant to acid attack as oxides. Nevertheless, extraction schemes have been developed [71–73]. Sizing experiments of the extracted inclusions have also been developed [74–76].

Analytical Methods for Oxygen and Sulfur

While the oxygen content of the steel can be calculated from the extracted inclusions, there are also a number of direct analytical methods for oxygen analysis. Neutron activation analysis [77] has been used, but vacuum and inert-gas fusion methods [78,79] are more commonly employed.

The analysis for sulfur content is a more routine matter, and a variety of techniques are available. Two methods, however, have widespread application: (1) The sample is burned in oxygen in an induction or resistance furnace, and the gaseous sulfur combustion products are measured by iodate or caustic titration. (2) An infrared Luft cell measures the sulfur dioxide combustion product. Since the solubility of sulfur in steel is quite low, measurement of the sulfur content is directly related to the sulfide content.

## MEASUREMENTS OF INCLUSIONS BY MICROSCOPIC METHODS

Since inclusions influence properties, both harmfully and beneficially, the metallurgist has been faced with the problem of obtaining quantitative information about inclusion content, size, shape and distribution in order to assess the magnitude of their influence. The previous sections described many methods that can be used to detect inclusions. While the quantifying nature of many of these methods were easily described, the microscopic methods are so diverse that this aspect will be developed separately at this point. The microscopic methods will be described under the following categories:
- Measurement of inclusion malleability
- Chart methods for assessing inclusion content
- Manual quantitative methods
- Automatic image analysis

Our review will focus on the practical aspects of such measurements, with particular attention to the influence of techniques on the statistical value of the measurements.

Inclusion Malleability

The malleability of inclusions [80–90] is an important parameter, because it is an aid to inclusion identification and because the shape of inclusions in wrought products has a direct influence on properties and behavior. Manganese sulfides and glassy manganese silicates deform readily during hot working, whereas the harder, more brittle refractory inclusions do not. In recent years, considerable effort has been put into developing globular sulfide shape to reduce loss of toughness and ductility in the transverse and through-thickness test directions.

Forgeng[30] classified inclusion types on the basis of their relative malleability during hot working (Table 3). Baker and Charles[80] showed that Type III manganese sulfides are highly deformable while Type I sulfides are significantly less deformable. The deformability of both types increases as the hot-working temperature decreases to the austenite-to-ferrite transformation temperature. Additionally, as the amount of hot working increases, the inclusion plasticity decreases. The malleability of Type II sulfides is difficult to determine because of the complexity of their initial shape. Above a certain temperature the plasticity of manganese silicate inclusions increases dramatically. This temperature is influenced by the composition of the silicates.

The analytical procedures were first introduced by Malkiewicz and Rudnik[89] and modified by Maunder and Charles[85] and others. If the initially spherical inclusion is deformed during hot working to an ellipsoid, the length-to-width ratio ($\lambda=b/a$) of the inclusion is constant for any section parallel to the major axis of the inclusion, i.e., in the hot-working direction. Measurements of the axial ratio of the inclusions are compared with the shape changes of the product from ingot to bar. The change in shape of the steel is calculated as follows[83]. If the cross-sectional area of the ingot is x and the cross-sectional area of the wrought product is y, then the reduction ratio of the steel product, h, is given by:

$$h = \frac{x^{3/2}}{y^{3/2}}$$

This ratio is compared with the measured aspect ratio, $\lambda$, of the inclusions to obtain a measure of the relative inclusion/steel plasticity.

Malkiewicz and Rudnik defined an index of deformability, $\nu$, for the inclusions as:

$$\nu = \frac{2 \log \lambda}{3 \log h}$$

## TABLE 3

### Classification of inclusions according to malleability by Forgeng[30]

Malleable Inclusions

Sulfides of iron, manganese and chromium
    [FeS, MnS, (Fe,Mn)S, (Fe,Cr,Mn)S]
Basic silicate glasses [$2(Fe,Mn)0 \cdot SiO_2$]

Slightly Malleable Inclusions

Silica-rich glasses [$(Fe,Mn)0 \cdot x SiO_2$]
Sulfides of titanium and zirconium
Siliceous chromites [$(Fe,Mn)0 \cdot Cr_2O_3 + SiO_2$]
Zircon [$ZrO_2 \cdot SiO_2$]

Nonmalleable Hard Refractory Inclusions

Silica [$SiO_2$]
Hercynite [$FeO \cdot Al_2O_3$]
Alumina [$Al_2O_3$]
Chromite (Fe,MnO $\cdot Cr_2O_3$]
Chromic Oxide [$Cr_2O_3$]
Ilmenite [$FeO \cdot TiO_2$]
Zirconium Oxide [$ZrO_2$]
Nitrides, carbides or carbonitrides of titanium, zirconium
    of vanadium
Vanadium spinel (FeO $\cdot V_2O_3$]

For nondeformable inclusions the value of the index is zero. If the inclusion elongates as much as the steel the index is equal to one. In general, the index of deformation is greater at low amounts of reductions than for high amounts and is lower for small inclusions than for large inclusions. Gove and Charles[88] showed that $\nu$ varies with the difference in hardness between the inclusion and matrix. Ekerot[86] claims that $\nu$ can be greater than one for silicate inclusions deformed just above their softening temperature.

Brunet and Bellot[90] modified the above formula for steel rolled to flat shapes. Two aspect ratios are calculated: $\lambda$ as above, and $\sigma = b/c$ where b is again the length of the inclusion in the longitudinal direction and c is the width of the inclusion in the transverse direction. The index of deformation is calculated as:

$$\nu = \frac{\log \lambda + \log \sigma}{3 \log h}$$

Baker and Charles[80] developed a method based on image analysis that eliminates the need for direct measurement of the aspect ratios. Using the Quantimet B, they measured the mean observed inclusion length, 2a', in the longitudinal direction and the number of inclusions per unit area, $N_A$. The mean projection per unit area of field, P, is calculated as:

$$P = (N_A) (2a')$$

The index of deformation, $\nu$, is equal to the inclusion true strain ($\epsilon$ for inclusion) divided by the true strain of the steel ($\epsilon$ for steel) where $\epsilon$ steel = 1nh. The inclusion true strain, $\epsilon$ for inclusion, is calculated as:

$$\epsilon_{inclusion} = \log a/a_o$$

where
     2a is the length of deformed inclusion, and
     $2a_o$ is the diameter of the spherical as-cast inclusion.
The apparent inclusion length, 2a', is related to the real length by:
$$2a' = 2a \sqrt{2/3}$$
Hence, the inclusion true strain is calculated as:

$$\epsilon_{inclusion} = \log a'/a'_o = \log P/P_o$$

where $P_o$ is the projection measurement from the cast steel.
The inclusion relative plasticity, $\nu$, is calculated as:

$$v = \frac{\log P - \log P_o}{\log h}$$

Standard Chart Methods

The oldest and most widely used microscopic methods for inclusion analysis are based on the use of comparison charts which pictorially illustrate different types, quantities and distributions of inclusions. In the application of these and indeed all microscopic methods the following factors[91] influence one's ability to accurately measure the inclusion content:
- Amount of hot working and inclusion plasticity
- Amount, location and orientation of samples
- Sample preparation
- Magnification and resolution

- Area examined per field and per sample
- Field selection
- Other considerations

Influence of Hot Working and Plasticity.  With standard charts, only a certain number of pictures can be used to characterize inclusions.  Since an important aspect of the schematic representation of the inclusions is shape, the malleability of the inclusions is clearly an important factor.  Hot working elongates the inclusions preferentially in the hot-working direction, but the amount of elongation varies according to type of inclusion, hot-working temperature, location within the material, and amount of deformation.  During rolling, the metal movement is greatest at the surface;  hence, inclusion elongation will vary through the section.

Influence of Sampling Scheme.  Since inclusion size and concentration vary through the product, all sampling procedures must standardize on test locations with respect to the location within the original ingot and across the section.  Clearly, the use of random sampling will require many more samples to characterize a given melt than samples chosen systematically.  Hence, samples are usually taken from the top and bottom and frequently from the middle of the first, middle and last ingot.  To assess inclusion malleability, the plane of polish is taken at the mid-thickness location and is parallel to the hot-working axis.  In practice, sampling is a compromise between the number of samples desired to achieve a satisfactory rating and the time and cost of analysis.  Such testing must be adequate to provide sound statistical evaluation but not so extensive as to be prohibitive.

Influence of Sample Preparation.  Sample preparation is critical.  To minimize statistical errors, sample preparation must be highly controlled.  Scratches, pits, comet-tailing and staining must be prevented, inclusions must be retained, and smearing must be avoided.  Automatic polishing devices prove to be quite valuable.  Low-nap cloths are recommended for minimizing relief.

If standard chart methods are used or if one is measuring inclusion or stringer length, sectioning errors can introduce serious errors.  Allmand and Coleman[92] studied the influence of the plane-of-polish orientation with respect to the hot-working axis on the accuracy of inclusion measurement.  They concluded that a minor misorientation of only six degrees from the true longitudinal plane introduce serious errors in counting inclusions.  Such deviations also influence length measurements.  Volume-fraction measurements, on the other hand, are insensitive to the plane of polish chosen.  As the orientation of the plane-of-polish changes, the apparent inclusion area, length, number of inclusions and distribution changes.

Influence of Magnification.  Most standard chart methods are established for observation at 100X magnification, which provides a large field of view and permits rapid assessment of a large area.  For steels with a high degree of freedom from inclusions, higher magnifications are necessary to identify the inclusions.  However, raising the magnification reduces the field of view, thus requiring a much greater effort to view the same total area.  For example, the field of view at 500X is only one twenty-fifth that at 100X.

Influence of Examination Area.  Generally, a polished microsample will have a surface area between about 0.5–1 sq. in.  If nine samples are characterized to represent a given heat, which may weigh anywhere from 20 to 300 tons, such a measurement would be based on a study of less than 10 sq. in. of surface.  Consequently, the major problem inherent in the method is the low probability that the inclusions observed on these surfaces are representative of all of the inclusions in the heat.

As the surface area examined increases, the probability of observing the largest inclusions in the heat increases. Indeed, studies have shown that as the number of samples, i.e., the test area increases, the inclusion ratings indicate poorer quality with respect to the inclusion content, especially if worst-field rating criteria are utilized. It is often useful to include samples in such studies that are chosen based on inclusions detected by other procedures such as ultrasonics, hot acid etching, and magnetic particle inspection. Such samples would provide greater assurance of detecting worst conditions.

Field Selection. In making inclusion measurements the operator has control of the field selection. If the entire surface is to be examined, the observation is begun at one corner of the specimen and contiguous fields of view are chosen until the entire surface is scanned. If only a portion of the area is scanned, one should move the stage in systematic increments without looking at the field of view while placing the chosen fields systematically but without bias throughout the entire sample surface. This practice will average any gradients within the specimen.

Identification of Inclusions. Numerous standard chart methods have been developed over the years. Most are designed to recognize the inclusion morphology typical of different types of indigenous inclusions. Exogenous inclusions and complex inclusions, i.e., one type intertwined with another, are not covered by the chart methods. Most schemes categorize inclusions as sulfides or oxides and generally further classify oxides by their plasticity: plastic (silicates), brittle (aluminates), or globular. If classification by such schemes is to be applicable to a wide variety of steel types, the observer must make decisions regarding the correlation of the actual inclusion shapes to those in the charts. Unfortunately, operators often base their decisions on somewhat subjective reasons, and there is little reproducibility between operators.

ASTM E45. In the United States, inclusion chart methods[10] are summarized in ASTM specification E45. While this standard includes recommended practices, it does not establish acceptance criteria. Such criteria are detailed in specifications for specific products. A number of ASTM specifications (Table 4) require inclusion assessment, but not all impose measurement criteria (Table 5).

The ASTM E45 charts are designed to determine the size, distribution, number and types of inclusions. Three charts are included: the JK (Jernkontoret) chart (Fig. 7) [93,94], the SAE chart[2] (Fig. 8), and a modified JK chart (Fig. 9) for vacuum-degassed and other steels with low inclusion contents. The original JK chart is used for the majority of steels, whereas the SAE chart is primarily for low-carbon steels intended for carburizing.

The JK chart (Plate I of E45) categorizes inclusions into: sulfides (Type A), aluminates (Type B), silicates (Type C), and globular (Type D). Classification is stated to be only by inclusion morphology, but most users consider composition as well. Each category is graded into thin and thick series, with five severities from 1 to 5. The modified JK chart (Plate III of E45) uses the same schemes, but the severity ratings are in five steps from ½ to 2½. Data are provided for inclusion Types A, B and C regarding the length and average width and thickness for each severity and the average diameter and number per field for each severity of Type D inclusions. The SAE chart (Plate II of E45) consists of two series of eight levels of severity. One series is referred to as slags or silicates (deformable inclusions), and the other as oxides (nondeformable inclusions). The silicate series is used to rate all malleable inclusions, regardless of composition.

## TABLE 4

### ASTM specification requiring inclusion assessment

Specifications Without Imposed Limits

| | |
|---|---|
| ASTM A108 | Standard Specification for Steel Bars, Carbon, Cold Finished, Standard Quality |
| ASTM A331 | Standard Specification for Steel Bars, Alloy, Cold-Finished |
| ANSI/ASTM A576 | Standard Specification for Steel Bars, Carbon, Hot-Rolled, Special Quality |
| ANSI/ASTM A579 | Standard Specification for Superstrength Alloy Steel Forgings |
| ANSI/ASTM A600 | Standard Specification for High-Speed Tool Steel |
| ANSI/ASTM A646 | Standard Specification for Premium Quality Alloy Steel Blooms and Billets for Aircraft and Aerospace Forgings |
| ANSI/ASTM A681 | Standard Specification for Alloy Tool Steels |
| ASTM A686 | Standard Specification for Carbon Tool Steels |
| ASTM A711 | Standard Specification for Carbon and Alloy Steel Blooms, Billets, and Slabs for Forging |

Specifications With Imposed Limits

| | |
|---|---|
| ASTM A295 | Standard Specification for High Carbon-Chromium Ball and Roller Bearing Steel |
| ASTM A485 | Standard Specification for High Hardenability Bearing Steels |
| ASTM A534 | Standard Specification for Carburized Steels for Anti-Friction Bearings |
| ANSI/ASTM A535 | Standard Specification for Special Quality Ball and Roller Bearing Steel |

## TABLE 5

### Summary of inclusion rating limits for specialty steels

| | Inclusion Type | | | | | | | |
|---|---|---|---|---|---|---|---|---|
| | A | | B | | C | | D | |
| Specification[1] | Thin | Thick. | Thin | Thick. | Thin | Thick. | Thin | Thick. |
| ASTM A295-70[2] | 2-1/2 | 1-1/2 | 2 | 1-1/2 | 2 | 1-1/2 | 1-1/2 | 1-1/2 |
| ASTM A485-75[2] | 2-1/2 | 1-1/2 | 2-1/2 | 1-1/2 | 2-1/2 | 2 | 2 | 1-1/2 |
| ASTM A534-76[3] | | | | | | | | |
| Electric Furnace | 3 | 2 | 3 | 2-1/2 | 2-1/2 | 1-1/2 | 2 | 1-1/2 |
| BOH or BOF | 3-1/2 | 2-1/2 | 3-1/2 | 3 | 3 | 2 | 2-1/2 | 2 |

Notes:  1)  Two thirds of all specimens and at least one from each ingot tested, as well as the average of all specimens, shall not exceed the listed ratings.
        2)  Use Plate III of ASTM E45 except for those inclusions exceeding a 2-1/2 rating. For these larger inclusions use Plate I of E45.
        3)  Use Plate I of ASTM E45 for ratings.

| | A | | B | | C | | D | |
|---|---|---|---|---|---|---|---|---|
| Specification | Thin | Thick. | Thin | Thick. | Thin | Thick. | Thin | Thick. |
| ANSI/ASTM A535-65 | 1-1/2 | 1 | 1-1/2 | 1 | 1-1/2 | 1 | 1-1/2 | 1 |
| Max. No. of Rateable Fields in any Sample | 8 | 1 | 3 | 1 | 3 | 1 | 5 | 1 |

Notes:  A rateable field is a field equal to or greater than 1 on the chart (Plate III). No individual D-type inclusions exceeding 0.001 in. (0.025 mm) in diameter, no more than three between 0.0005 and 0.001 in. (0.013 and 0.025 mm). No individual A-type inclusions exceeding 0.0075 in. (0.19 mm) in length.

G.F. VANDER VOORT

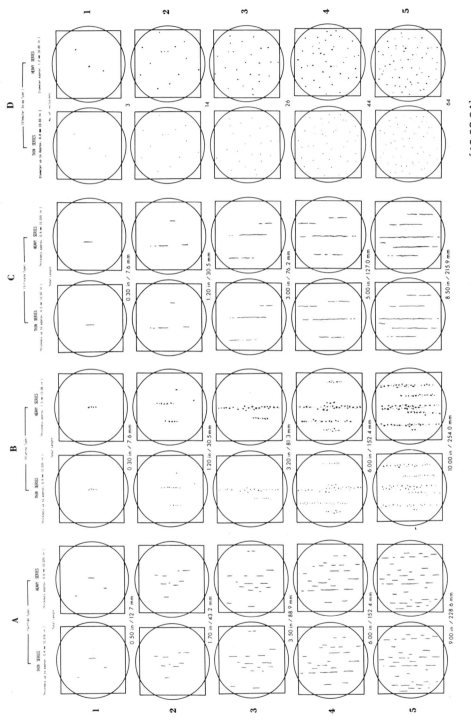

Fig. 7. The Jernkontoret chart for rating inclusion content, Plate 1 of ASTM E45[10,93,94].

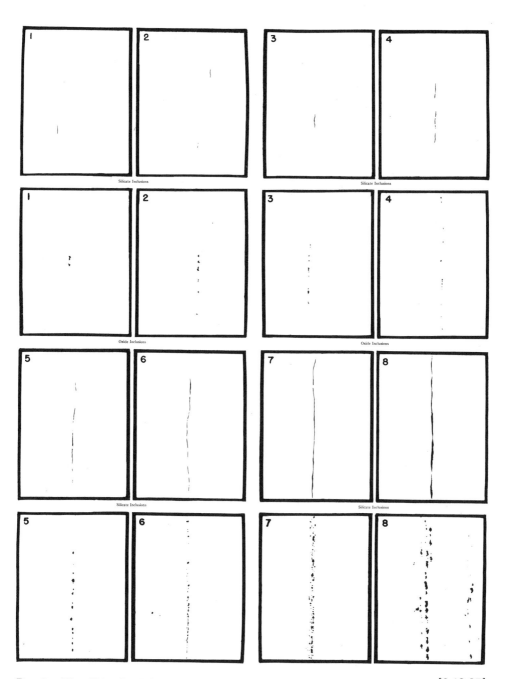

Fig. 8. The "Slag-Oxide" chart developed by Walker, Plate II of ASTM E45[2,10,97].

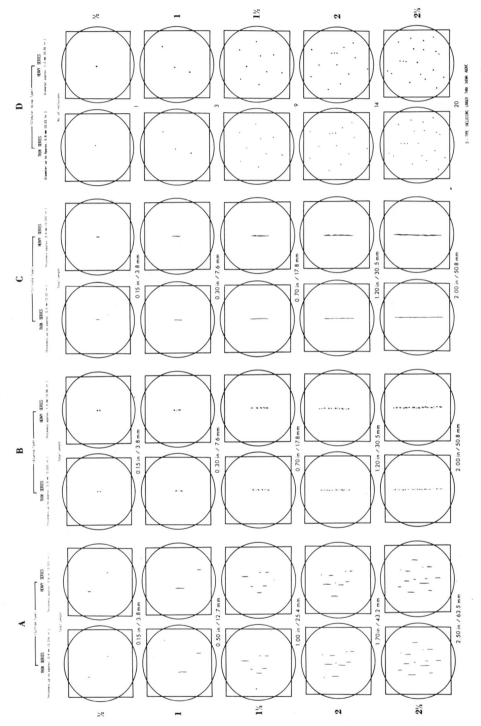

Fig. 9.   Modified Jernkontoret chart developed by ASTM committee E4–09, Plate III of ASTM E45[10].

ASTM E45 contains four microscopic methods for evaluating inclusion content. Three of these methods utilize the above chart methods. Method A uses the JK chart, method C the SAE chart, and method D uses the modified JK chart. Method B will be discussed later.

Other Chart Methods. Standard chart methods have been developed in many countries. Perhaps the earliest chart is one developed by an unidentified Swedish steelmaker (Fig. 10), which was reproduced by Benedicts and Lofquist[21] in their classic text on inclusions published in 1931.

In England the chart procedure developed by Bolsover[95] was adopted by S. Fox & Co.[96] and has since been referred to as the Fox Chart (Fig. 11). Other early chart methods include the previously mentioned JK method developed in Sweden, the Chevrolet Slag-Oxide chart (ASTM E45, Plate II) developed by Walker[2,97], and the Diergarten Chart (Fig. 12) developed in Germany[3,4]. In recent years chart methods have been developed in Germany[98,99] for specialty steels* (Fig. 13) and for free-machining steels[100] (Fig. 14). The Italian Standard[101] utilizes either the JK Chart or the Diergarten Chart for inclusion assessment. In Japan[102] a chart method was adopted in 1956 by the Japan Society for Promotion of Science but was replaced by methods that did not utilize standard charts when their specification JIS–G–0555 was adopted[103]. Two chart methods have been developed in Russia[104,105]. The first method, GOST–1778, is intended for a wide variety of steels (Fig. 15), while the second method, GOST–801, is intended for bearing steels (Fig. 16). Other Russian chart-based specifications include GOST–841 for ball-bearing steels and GOST–803 for low-carbon rimming steels. Polish Standard PN–67/H–84 041 for ball and roller bearing steels is a refinement of the Russian GOST–801 Chart.

Chart methods not incorporated into national specifications include those of Alisanova[106], Nulk[107], Schreiber and Ziehm[108], Ruhl[109], Scheil[110], and Malinochka and Moiseev[111]. This last reference is rather interesting in that it correlates inclusion volume fraction with the severity ratings for randomly dispersed and clustered inclusions (Fig. 17). Inclusion charts have also been developed by a number of companies. Perhaps the best known is the FAG Chart[112] shown in Fig. 18. A chart developed by Breda Siderurgica Milano of Italy for rating sulfides in free-machining steels is shown in Reference 113.

Application of Chart Methods. Chart methods are used in two ways. In the simpler procedure, each sample used for evaluating the inclusion content of a heat is surveyed and the worst field of each inclusion chart type is recorded. After all of the samples for the heat are evaluated, the averages of the worst fields of each inclusion type are calculated to represent the inclusion content of the heat. The second method is more quantitative. The operator records a rating for every field examined and summarizes the number of fields in each severity rating classification and for each inclusion type. These data may be further analyzed by multiplying the number in each category by a weighting factor to obtain an index number representing the relative inclusion content of the sample. Some specifications include such procedures, while others, such as ASTM E45, do not. In such cases, manufacturers establish their own weighting factors.

Evaluation of Chart Methods. Because of the wide use of standard chart methods and the variability of the criteria on which they are based, the factors that influence the value of inclusion assessment by chart comparison require thorough evaluation. These factors include:

_____
* The chart shown in Fig. 13 replaced an earlier version, VDEh 1570–61.

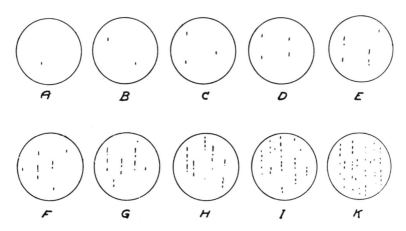

Fig. 10. Oldest known inclusion chart developed by an unidentified Swedish steel-maker; published by Benedicts and Lofquist[21].

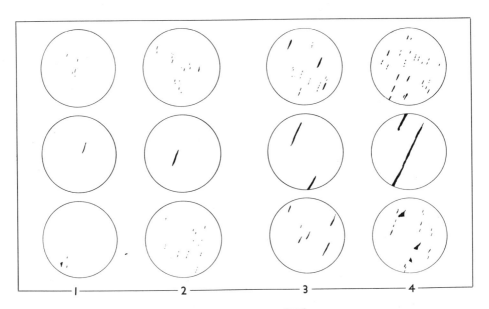

Fig. 11. The "Fox" chart developed by Bolsover[95] and adopted by S. Fox & Co. [96].

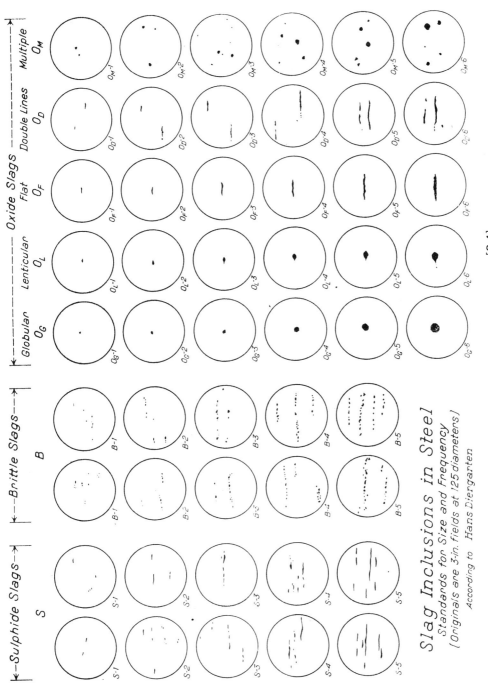

Fig. 12. Inclusion chart developed by Diergarten[3,4] in Germany.

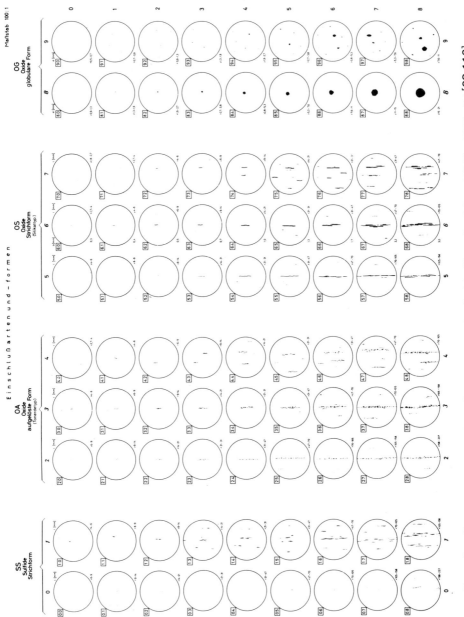

Fig. 13. German inclusion chart for rating steels, Stahl–Eisen Prufblatt VDEh 1570–71 [98,116].

Mikroskopische Prüfung von Automatenstählen auf sulfidische nichtmetallische Einschlüsse mit Bildreihen (Abbildungsmaßstab 100:1)

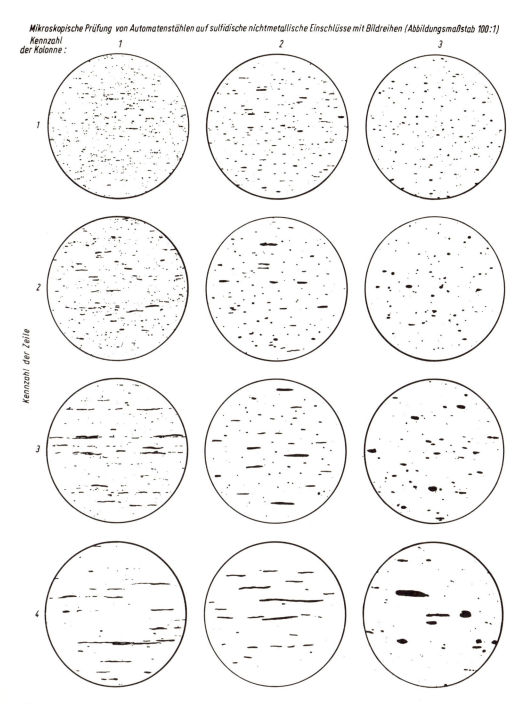

Fig. 14.  German inclusion chart for rating free-machining steels, Stahl—Eisen Pruf-blatt VDEh 1572[99,100].

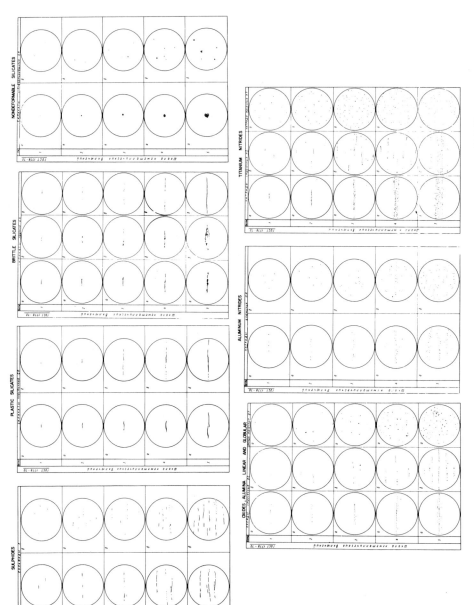

Fig. 15.   Russian inclusion chart for rating steels, GOST 1778–70[104].

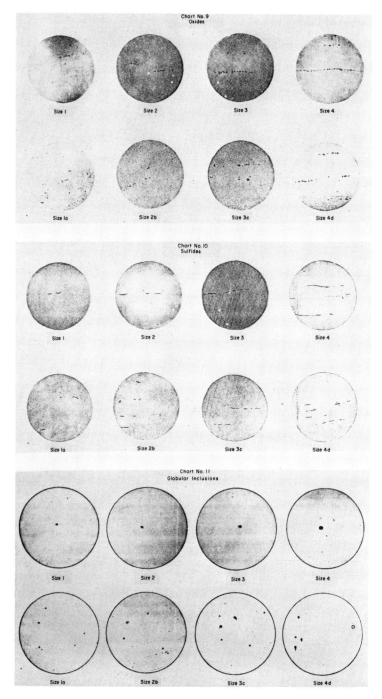

Fig. 16. Russian inclusion chart for rating ball and roller bearing steels, GOST 801—60[105].

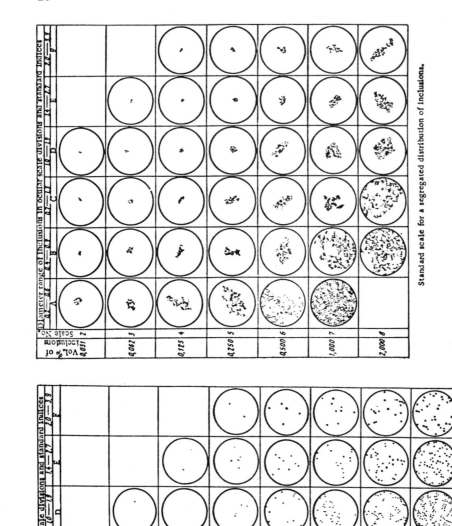

Fig. 17.   Inclusion chart developed by Malinochka and Moiseev[111] for rimming steels.

Guideline Series for Judging Nonmetallic Inclusions in Roller Bearing Steels

Magnification 100:1

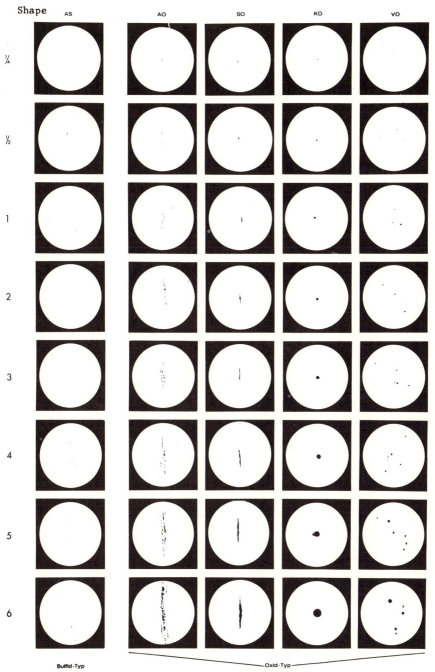

Fig. 18. The FAG inclusion chart adopted by Georg Schafer & Co.[112].

- Inclusion-picture grading
- Relevance of inclusions depicted
- Variation and validity of procedures
- Methods in use for enhancing chart ratings
- Statistical accuracy of chart ratings
- Correlation of chart ratings with other assessment methods

Advantages and Disadvantages of Chart Methods. Inclusion chart methods are relatively simple to apply, and an analysis is performed in a reasonably short time. However, they do suffer from a number of disadvantages that seriously affect the accuracy and reproducibility of the measurements. One of the best studies of different inclusion rating methods was conducted by Eckstein and Schneider (Ref. 114). In an evaluation of steel made from four heats they compared the effectiveness of macroscopic, chemical, and microscopic methods. The microscopic method was the Diergarten chart. Of the seven methods they evaluated, the microscopic method did not give a statistically certain agreement (i.e., 95% confidence level) with any of the other methods. They found that the step-down magnetic particle test gave the best agreement with the other tests. They concluded that the surface area evaluated with the Diergarten Chart was too small, i.e., there was a low probability of detecting the larger inclusions present.

As stated in the ASTM E45 Standard, the severity ratings for Types A, B and C inclusions are based on inclusion length, and the thin-versus-thick subdivision is based on inclusion width. Inclusion length and width are influenced by inclusion type, which is accounted for by the specification and by the amount of hot working. Given the fact that the amount of hot working varies widely among different products, the degree of hot working must be fixed for a given chart because all possibilities cannot be represented. The E45 charts were established to be representative of the amount of inclusion elongation developed during rolling of an ingot to a 4X4-inch billet. For many products the steelmaker can obtain samples during the processing when the product is at or close to this size. However, such sampling is not possible in the case of castings, large forgings, and thick plates. For the user who obtains a finished product with a markedly different amount of reduction, the inability to obtain samples with the correct amount of hot reduction presents a serious problem because inclusion appearance does not conform to chart pictures. Many of the problems inherent in inclusion ratings according to the ASTM chart methods were reviewed by Hoo[115].

While inclusion recognition under E45 is governed by morphology alone, it would be unwise to classify silicates as sulfides or vice versa if the form of an inclusion does not conform to the correct type. Most users prefer to classify inclusions on the basis of chemical type first and shape second. This approach requires the ability to recognize basic inclusion types by color and other differences. For the basic types such recognition should be an easy matter, but it is the writer's experience that errors are made in differentiating among sulfides, silicates and aluminates. To the eye, manganese sulfides appear grey, whereas silicates are black and glassy. The JK charts represent all inclusions with black ink of the same density. Only in the case of the German standard VDEh 1570 and the FAG Chart are the sulfides depicted as being lighter than the silicates. For steels with a low inclusion content it is often impossible to identify specific inclusions at 100X; hence, the observer must switch to a higher magnification for proper identification before classification at 100X. Along with color, the shape of the inclusion tips is a good indicator of type. Sulfide tips are more rounded than silicate tips, which are relatively sharp.

In the as-cast condition, the shape of aluminates is quite complex. Since they are very hard and brittle, they do not deform during rolling but are broken up and strung out in the rolling direction. Estimation of the length of aluminate stringers becomes somewhat subjective. According to E45, if the separation between particles in an apparent inclusion stringer is less than 0.005 inch, the inclusion is considered to be part of the stringer. According to some investigators, if the gap between adjacent particles is less than the size of the particles, these should be considered to be part of the same stringer.

In the thin-and-thick classification scheme all of the inclusions in a given thin series are less than a certain thickness while all of the inclusions in a given thick series are above a certain thickness. In actual practice, however, all of the inclusions in a given field do not conform to this idealized classification, and the operator is faced with the need to make subjective decisions. Since the plane of polish does not intersect all inclusions along their centerline, a variety of thicknesses can be expected.

The classification of complex inclusions is difficult, because they do not conform to any picture series. It is not unusual to observe sulfide inclusions that have precipitated on a less deformable oxide inclusion. One inclusion type can completely or almost completely surround another, or particles of one type may be at the tips of another.

During the development of the new VDEh 1570–71 Chart, Barteld and Stanz [116] reviewed the chief existing standards: American (ASTM E45), German (VDEh 1570–61), Russian (GOST–1778 and 801), British (Fox Chart), and the FAG Chart. They pointed out some of the problems associated with these charts. For example, they noted that the original JK Chart (Plate I of ASTM E45) was developed using steels made in electric furnaces lined with acid rather than basic brick. The production of acid electric furnace steels is negligible today. Unfortunately, there are fundamental differences between the inclusions in steels melted in acid-lined versus basic-lined furnaces. In particular, the sulfides (Type A) in acid furnaces are depicted in this chart as numerous short lines (referred to as "rain pictures"). This error was incorporated with the modified JK Chart first introduced in the 1963 revision of E45. The appearance of the other inclusion types (B,C and D) are also somewhat different in acid-lined electric furnace steels.

Measurement of ASTM E45 Inclusion Rating Charts. The author measured the inclusions presented on Plates I, II and III of ASTM E45 (see Figs. 7, 8 and 9), and an examination of the resulting information reveals the formidable problems in developing and using such charts. We will consider Plates I (JK Chart) and III (Modified JK Chart) together, since the modified chart should merely be a refinement and extension of the original chart. Because the two charts have overlapping severity classifications, the descriptive pictures should be mutually compatible. However, as will be shown, they are not. Table 6 contains the numerical data from Plates I and III on which classifications are based. Table 6 should be referred to during the following discussion of the measured values.

The volume fraction (%) of inclusions present in each field given on Plates I and III was measured using the Leitz Texture Analyzer System. Results are presented in Table 7 and are shown graphically in Figs. 19–22. The data points for the Plate I measurements show a trend that is generally linear on the log-log plot, although a number of points deviate from the trend. The Plate III data show a similar trend, but the ½ severity points deviate significantly in 5 of 8 cases. If the two charts are

## TABLE 6

### Numerical data given on Plates I and III of ASTM E45.

All Dimensions in mm at 100X

| | | Inclusion Type | | | | | | | | | |
|---|---|---|---|---|---|---|---|---|---|---|---|
| | Type A | | | Type B | | | Type C | | | Type D | |
| | | Thickness | | | Thickness | | | Thickness | | | Diameter | |
| Severity | Length | Thin | Thick. | Length | Thin | Thick. | Length | Thin | Thick. | Number | Thin | Thick. |
| | | | | Plate I (JK Chart) | | | | | | | | |
| 1 | 12.7 | ≤.4 | ∿.6 | 7.6 | ≤.9 | ∿1.5 | 7.6 | ≤.5 | ∿.9 | 3 | ≤.8 | ∿1.2 |
| 2 | 43.2 | ≤.4 | ∿.6 | 30.5 | ≤.9 | ∿1.5 | 30.5 | ≤.5 | ∿.9 | 14 | ≤.8 | ∿1.2 |
| 3 | 88.9 | ≤.4 | ∿.6 | 81.3 | ≤.9 | ∿1.5 | 76.2 | ≤.5 | ∿.9 | 26 | ≤.8 | ∿1.2 |
| 4 | 152.4 | ≤.4 | ∿.6 | 152.4 | ≤.9 | ∿1.5 | 127.0 | ≤.5 | ∿.9 | 44 | ≤.8 | ∿1.2 |
| 5 | 228.6 | ≤.4 | ∿.6 | 254.0 | ≤.9 | ∿1.5 | 215.9 | ≤.5 | ∿.9 | 64 | ≤.8 | ∿1.2 |
| | | | | Plate III (Mod. JK Chart) | | | | | | | | |
| 1/2 | 3.8 | ≤.4 | ∿.6 | 3.8 | ≤.9 | ∿1.5 | 3.8 | ≤.5 | ∿.9 | 1 | ≤.8 | ∿1.2 |
| 1 | 12.7 | ≤.4 | ∿.6 | 7.6 | ≤.9 | ∿1.5 | 7.6 | ≤.5 | ∿.9 | 3 | ≤.8 | ∿1.2 |
| 1-1/2 | 25.4 | ≤.4 | ∿.6 | 17.8 | ≤.9 | ∿1.5 | 17.8 | ≤.5 | ∿.9 | 9 | ≤.8 | ∿1.2 |
| 2 | 43.2 | ≤.4 | ∿.6 | 30.5 | ≤.9 | ∿1.5 | 30.5 | ≤.5 | ∿.9 | 14 | ≤.8 | ∿1.2 |
| 2-1/2 | 63.5 | ≤.4 | ∿.6 | 50.8 | ≤.9 | ∿1.5 | 50.8 | ≤.5 | ∿.9 | 20 | ≤.8 | ∿1.2 |

## TABLE 7

### Volume fraction (%) of inclusions in ASTM E45.

| | Inclusion Type | | | | | | | |
|---|---|---|---|---|---|---|---|---|
| Severity | A | | B | | C | | D | |
| | Thin | Thick. | Thin | Thick. | Thin | Thick. | Thin | Thick. |
| Plate I | | | | | | | | |
| 1 | .137 | .143 | .08 | .186 | .176 | .18 | .083 | .148 |
| 2 | .36 | .477 | .274 | .65 | .34 | .404 | .291 | .517 |
| 3 | .809 | 1.09 | .615 | 1.77 | .58 | .692 | .457 | .911 |
| 4 | 1.84 | 2.313 | 1.191 | 3.08 | .96 | 1.18 | .704 | 1.45 |
| 5 | 2.67 | 3.73 | 2.81 | 7.0 | 1.73 | 2.33 | 1.0 | 2.09 |
| Plate II | | | | | | | | |
| 1/2 | .023 | .035 | .036 | .068 | .032 | .042 | .0092 | .025 |
| 1 | .069 | .09 | .069 | .101 | .062 | .089 | .032 | .053 |
| 1-1/2 | .085 | .237 | .099 | .204 | .161 | .188 | .092 | .161 |
| 2 | .192 | .325 | .155 | .312 | .281 | .327 | .151 | .262 |
| 2-1/2 | .405 | .505 | .214 | .354 | .517 | .551 | .188 | .368 |

to be rendered compatible, the lines representing the best fit of the data should over-lap with the same slope. None of the lines for Plates I and III, thin or thick, and Types A, B, C or D overlap. Only for the A and D inclusions are the trends similar, that is, the lines are parallel. Severity ratings 4 and 5 are typical of very high inclusion contents. For example, these two ratings for sulfides are typical for heavily resulfurized free-machining steels. For all cases there is a substantial difference between the inclusions with severity ratings in the overlapping region represented by severities ratings 1 and 2.

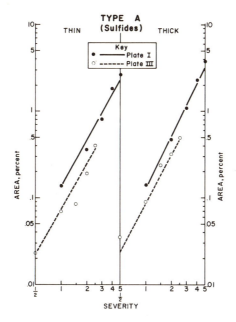

Fig. 19.  Measurement of volume fraction of thin and thick sulfide (Type A) inclusions as a function of severity, Plates I and III, ASTM E45.

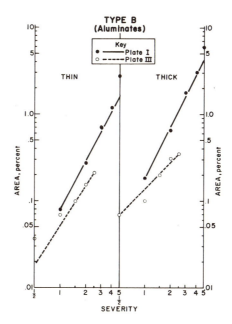

Fig. 20.  Measurement of volume fraction of thin and thick aluminate (Type B) inclusions as a function of severity, Plates I and III, ASTM E45.

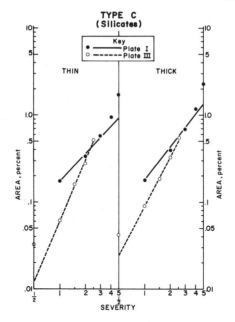

Fig. 21.  Measurement of volume fraction of thin and thick silicate (Type C) inclusions as a function of severity, Plates I and III, ASTM E45.

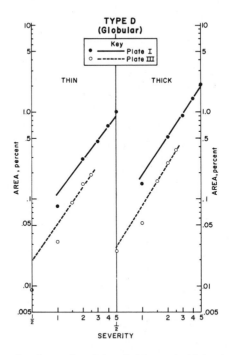

Fig. 22.  Measurement of volume fraction of thin and thick globular (Type D) inclusions as a function of severity, Plates I and III, ASTM E45.

The inclusions in each picture were counted and Table 8 presents these data. The number of inclusions per square millimeter of surface area examined were calculated for 1X magnification (Table 9). The data in Tables 8 and 9 are plotted in Figs. 23–26. Except for the B1, C5 and D1 fields, the Plate I data fall along a straight line in the plot. The Plate III data do not fit the straight-line trend. For Types A and D, the two severity ratings by both methods agree well but the ½ and 1 ratings do not. For Types B and C, the two methods are not in agreement. Type C inclusions are represented in Plate III by a single inclusion per field, regardless of the severity rating. This fact is noted in E45, but no reasons are given. The numbers of Type D inclusions per field (Table 8) agree well with the listed data (Table 6), except for D–4 thick and D–5 thin, each of which is missing one inclusion.

## TABLE 8

Number of inclusions in each chart picture, ASTM E45.

| Severity | Inclusion Type | | | | | | | |
|---|---|---|---|---|---|---|---|---|
| | A | | B | | C | | D | |
| | Thin | Thick. | Thin | Thick. | Thin | Thick. | Thin | Thick. |
| Plate I | | | | | | | | |
| 1 | 3 | 3 | 6 | 5 | 1 | 1 | 3 | 3 |
| 2 | 9 | 9 | 17 | 14 | 5 | 5 | 14 | 14 |
| 3 | 16 | 16 | 43 | 40 | 10 | 8 | 26 | 26 |
| 4 | 22 | 22 | 86 | 93 | 14 | 13 | 44 | 43 |
| 5 | 37 | 37 | 150 | 147 | 13 | 14 | 63 | 64 |
| Plate III | | | | | | | | |
| 1/2 | 2 | 2 | 2 | 2 | 1 | 1 | 1 | 1 |
| 1 | 2 | 2 | 4 | 4 | 1 | 1 | 3 | 3 |
| 1-1/2 | 6 | 6 | 14 | 14 | 1 | 1 | 9 | 9 |
| 2 | 9 | 9 | 15 | 17 | 1 | 1 | 14 | 14 |
| 2-1/2 | 13 | 13 | 29 | 25 | 1 | 1 | 20 | 20 |

## TABLE 9

Number of inclusions per mm$^2$ at 1X, ASTM E45

| Severity | Inclusion Type | | | | | | | |
|---|---|---|---|---|---|---|---|---|
| | A | | B | | C | | D | |
| | Thin | Thick. | Thin | Thick. | Thin | Thick. | Thin | Thick. |
| Plate I | | | | | | | | |
| 1 | 6.3 | 6.3 | 12.6 | 10.5 | 2.1 | 2.1 | 6.3 | 6.3 |
| 2 | 18.9 | 18.9 | 35.6 | 29.4 | 10.5 | 10.5 | 29.4 | 29.4 |
| 3 | 33.5 | 33.5 | 90.1 | 83.9 | 21 | 16.8 | 54.5 | 54.5 |
| 4 | 46.1 | 46.1 | 180.3 | 195 | 29.4 | 27.3 | 92.2 | 90.1 |
| 5 | 77.6 | 77.6 | 314.5 | 308.2 | 27.3 | 29.4 | 132.1 | 134.2 |
| Plate III | | | | | | | | |
| 1/2 | 4.2 | 4.2 | 4.2 | 4.2 | 2.1 | 2.1 | 2.1 | 2.1 |
| 1 | 4.2 | 4.2 | 8.4 | 8.4 | 2.1 | 2.1 | 6.3 | 6.3 |
| 1-1/2 | 12.6 | 12.6 | 29.4 | 29.4 | 2.1 | 2.1 | 18.9 | 18.9 |
| 2 | 18.9 | 18.9 | 31.4 | 35.6 | 2.1 | 2.1 | 29.4 | 29.4 |
| 2-1/2 | 27.2 | 27.2 | 60.8 | 52.4 | 2.1 | 2.1 | 41.9 | 41.9 |

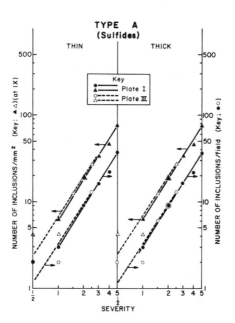

Fig. 23.  Number of thin and thick sulfide inclusions per field and per mm² as a function of severity, Plates I and III, ASTM E45.

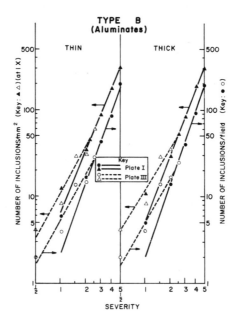

Fig. 24.  Number of thin and thick aluminate inclusions per field and per mm² as a function of severity, Plates I and III, ASTM E45.

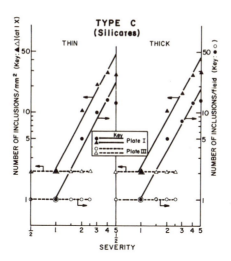

Fig. 25.  Number of thin and thick silicate inclusions per field and per mm² as a function of severity, Plates I and III, ASTM E45.

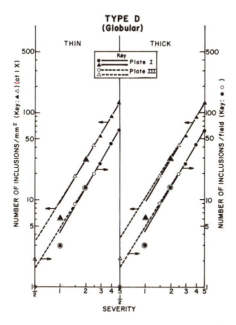

Fig. 26.  Number of thin and thick globular inclusions per field and per mm² as a function of severity, Plates I and III, ASTM E45.

The total inclusion length per field for inclusions of Types A, B and C is given in Table 10. The data have been calculated for 1X magnification. In comparing these data with the listed values in Table 6, multiply by 100 to give the total length at 100X. The total inclusion length of the Type A inclusions  agrees reasonably well with the data in Table 6. Note, however, that there is some difference in total length between equal-severity thin and thick chart fields. The total inclusion lengths of the B and C inclusions do not agree with the Table 6 data. The listed B and C inclusion lengths are based on total stringer lengths. Table 11 lists the average inclusion lengths for Types A, B and C. The data show that the average inclusion length measurement for Type A inclusions is relatively constant as severity increases. The average size of Type B inclusions decreases as the severity rating increases, a feature that appears to be unnatural. As would be expected, there is no relationship between the average length of Type C inclusions presented on the two charts. Figure 27 is a plot of the average and total inclusion lengths for Type A inclusions on Plates I and III of E45. Note the excellent agreement in total inclusion length between Plates I and III. The average inclusion lengths, except for the A1/2 and A1 measurements of Plate III, are in good agreement. The average diameter of the Type D inclusions (Table 12) agrees reasonably well with the stated chart values.

## TABLE 10

### Total inclusion length (1X magnification), E45

| Severity | Inclusion Type | | | | | |
| | A | | B | | C | |
| | Thin | Thick. | Thin | Thick. | Thin | Thick. |
| Plate I | | | | | | |
| 1 | .132 | .132 | .065 | .06 | .086 | .095 |
| 2 | .437 | .436 | .145 | .226 | .259 | .265 |
| 3 | .886 | .89 | .309 | .501 | .668 | .657 |
| 4 | 1.458 | 1.483 | .649 | .996 | 1.098 | 1.051 |
| 5 | 2.227 | 2.208 | .979 | 1.668 | 1.854 | 1.961 |
| Plate III | | | | | | |
| 1/2 | .038 | .045 | .023 | .031 | .041 | .041 |
| 1 | .145 | .142 | .053 | .066 | .085 | .091 |
| 1-1/2 | .261 | .24 | .10 | .134 | .19 | .189 |
| 2 | .414 | .385 | .187 | .186 | .31 | .311 |
| 2-1/2 | .619 | .616 | .242 | .357 | .512 | .512 |

Note:  Multiply by 100 to obtain total length at 100X.

The numbers of B and C inclusion stringers were counted, and their total and average lengths were determined as listed in Table 13. The data generally agree well with the listed data. Only one inclusion stringer per field of the B and C inclusions has been presented in Plate III. Hence, the two severity ratings by the two charts are not compatible. Figures 28 and 29 present the data graphically. Except for the 1/2 severity ratings, there is good agreement between the two charts for B and C inclusion stringer lengths. As would be expected, there is no relationship between the average stringer lengths as given by the two charts.

## TABLE 11

### Calculation of average inclusion length (1X), ASTM E45

| Severity | Inclusion Type | | | | | |
|---|---|---|---|---|---|---|
| | A | | B | | C | |
| | Thin | Thick. | Thin | Thick. | Thin | Thick. |
| Plate I | | | | | | |
| 1 | .044 | .044 | .0108 | .012 | .086 | .095 |
| 2 | .0486 | .048 | .0085 | .016 | .052 | .053 |
| 3 | .0554 | .0556 | .0072 | .0125 | .067 | .082 |
| 4 | .0663 | .0674 | .0075 | .0107 | .078 | .081 |
| 5 | .0602 | .0597 | .0065 | .0113 | .143 | .14 |
| Plate III | | | | | | |
| 1/2 | .019 | .023 | .0115 | .0155 | .041 | .041 |
| 1 | .073 | .071 | .0133 | .0165 | .085 | .091 |
| 1-1/2 | .044 | .04 | .007 | .0096 | .19 | .19 |
| 2 | .046 | .043 | .013 | .0109 | .31 | .31 |
| 2-1/2 | .048 | .047 | .008 | .0143 | .512 | .512 |

NOTE:  Multiply by 100 to obtain average length at 100X.

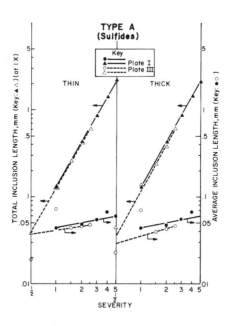

Fig. 27.  Total and average inclusion length of thin and thick sulfide inclusions as a function of severity, Plates I and III, ASTM E45.

## TABLE 12

### Measurement of average diameter of D-type inclusions (1X), ASTM E45

| Severity | D Inclusions | |
|---|---|---|
| | Thin | Thick. |
| Plate I | | |
| 1 | .0087 | .0157 |
| 2 | .0077 | .012 |
| 3 | .007 | .0122 |
| 4 | .0075 | .0129 |
| 5 | .0077 | .0116 |
| Plate III | | |
| 1/2 | .01 | .018 |
| 1 | .01 | .015 |
| 1-1/2 | .0096 | .012 |
| 2 | .0099 | .012 |
| 2-1/2 | .010 | .012 |

NOTE: Multiply by 100 to obtain average diameter at 100X.

## TABLE 13

### Measurement of number and length of B- and C-type stringers (1X), ASTM E45

| Severity | Number of Stringers | | | | Total Stringer Length (mm) | | | | Avg. Stringer Length (mm) | | | |
|---|---|---|---|---|---|---|---|---|---|---|---|---|
| | B | | C | | B | | C | | B | | C | |
| | Thin | Thick. | Thin | Thick. | Thin | Thick. | Thin | Thick. | Thin | Thick. | Thin | Thick. |
| Plate I | | | | | | | | | | | | |
| 1 | 1 | 1 | 1 | 1 | .098 | .096 | .086 | .095 | .098 | .096 | .086 | .095 |
| 2 | 2 | 2 | 1 | 1 | .315 | .284 | .194 | .193 | .158 | .142 | .194 | .193 |
| 3 | 2 | 2 | 3 | 3 | .806 | .822 | .764 | .766 | .403 | .411 | .255 | .255 |
| 4 | 4 | 4 | 4 | 4 | 1.516 | 1.522 | 1.265 | 1.282 | .379 | .381 | .316 | .321 |
| 5 | 5 | 5 | 5 | 5 | 2.211 | 2.557 | 2.063 | 2.073 | .442 | .511 | .413 | .415 |
| Plate III | | | | | | | | | | | | |
| 1/2 | 1 | 1 | 1 | 1 | .038 | .042 | .041 | .041 | .038 | .042 | .041 | .041 |
| 1 | 1 | 1 | 1 | 1 | .086 | .083 | .085 | .091 | .086 | .083 | .085 | .091 |
| 1-1/2 | 1 | 1 | 1 | 1 | .193 | .194 | .19 | .189 | .193 | .194 | .19 | .189 |
| 2 | 1 | 1 | 1 | 1 | .315 | .323 | .31 | .311 | .315 | .323 | .31 | .311 |
| 2-1/2 | 1 | 1 | 1 | 1 | .529 | .513 | .512 | .512 | .529 | .513 | .512 | .512 |

NOTE: Multiply by 100 to obtain total or average length at 100X.

Overall, the above data demonstrate that the two charts are different to begin with and that the rating of the same sample by both charts will produce different results. Although the data presented above could be used to relate inclusion measurement information from automatic image analysis to JK ratings, the results would still not be accurate, given the differences between the two charts. The data also show that only certain measured parameters can be related to JK numbers. Even more of a problem is that the JK charts are based on acid electric furnace steels

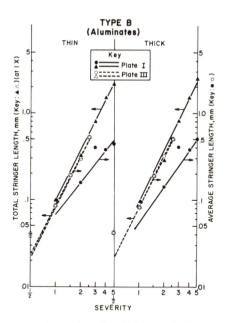

Fig. 28. Total and average stringer length of thin and thick aluminate inclusions as a function of severity, Plates I and III, ASTM E45.

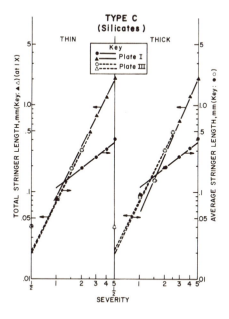

Fig. 29. Total and average stringer length of thin and thick silicate inclusions as a function of severity, Plates I and III, ASTM E45.

with inclusions different from those in most steels made today. If the worst field of each type were rated by either Plate I or Plate III, the histogramming of parameters measured by image analysis would reveal similar worst-field conditions. However, the image analysis method is more likely to detect the worst field present in a given sample, because this method is more objective and because a greater surface area can be examined.

The SAE inclusion chart in Fig. 8 (Plate II of E45) was also subjected to the same type of critical analysis as were Plates I and III. This chart does not list any values for inclusion size or number. Our measured data are presented in Table 14. Note that the volume fraction of inclusions varies from 0.014 to 1.04% for the silicates (deformable inclusions) and from 0.07 to 1.49% for the oxides (nondeformable inclusions) as the severity increases from 1 to 8. This gradation is more practical for rating most steels than the volume fraction gradation in Plate I. Figure 30 is a plot of area fraction measurement versus severity rating. The silicate gradation is fairly good, but the oxide gradation is quite poor. Figure 31 is a plot of the number of inclusions per field versus severity rating. The S7, S8 and O7 fields are out of line with the remaining fields. Figure 32 is a plot of total inclusion length versus severity rating. Except for S1, the agreement is not too bad for the silicate series. The oxide series has a very poor agreement between total inclusion length and severity. Figure 33 shows the poor relationship between average inclusion length and severity rating for both categories. As seen in Fig. 34, the total inclusion stringer length agrees better than average inclusion length with severity for the oxide series but gives a poor agreement for the silicate series. These results show that the SAE Chart is poorly graded in comparison with the JK and modified JK Charts.

## TABLE 14

### Measurements of inclusions — ASTM E 45 Plate II, Method C

| Severity | Silicates | | | | | | Oxides | | | | | |
|---|---|---|---|---|---|---|---|---|---|---|---|---|
| | No. of Incl. | No./mm$^2$ | Total Length (mm) | Avg. Length (mm) | Stringer Length (mm) | Area (%) | No. of Incl. | No./mm$^2$ | Total Length (mm) | Avg. Length (mm) | Stringer Length (mm) | Area (%) |
| 1 | 1 | 1.22 | .055 | .055 | .055 | .014 | 2 | 2.45 | .044 | .022 | .072 | .07 |
| 2 | 3 | 3.67 | .076 | .025 | .055 | .0262 | 8 | 9.8 | .104 | .013 | .28 | .114 |
| 3 | 4 | 4.9 | .112 | .028 | .09 | .0538 | 11 | 13.47 | .105 | .0095 | .435 | .076 |
| 4 | 9 | 11.02 | .277 | .035 | .52 | .124 | 13 | 15.92 | .093 | .0072 | .365 | .30 |
| 5 | 14 | 17.15 | .407 | .029 | .73 | .187 | 19 | 23.27 | .217 | .0114 | .723 | .264 |
| 6 | 12 | 14.7 | .678 | .057 | .86 | .289 | 44 | 53.9 | .378 | .0086 | .871 | .308 |
| 7 | 2 | 2.45 | .99 | .495 | 1.015 | .552 | 104 | 127.39 | .755 | .0073 | 1.04 | .742 |
| 8 | 3 | 3.67 | 1.02 | .341 | 1.05 | 1.04 | 64 | 78.39 | 1.108 | .0173 | 1.94 | 1.49 |

* All measurements in true length, i.e., chart measurement divided by 100.

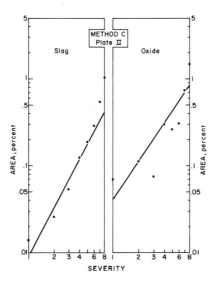

Fig. 30.  Measurement of volume fraction of slag and oxide type inclusions as a function of severity, Plate II, ASTM E45.

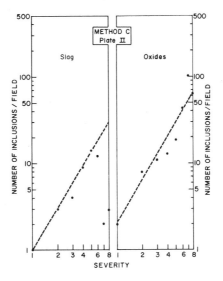

Fig. 31.  Number of slag and oxide type inclusions per field and per mm$^2$ as a function of severity, Plate II, ASTM E45.

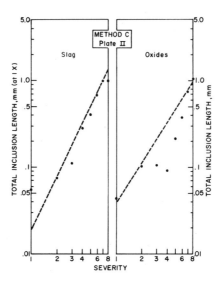

Fig. 32. Total slag and oxide type inclusion length as a function of severity, Plate II, ASTM E45.

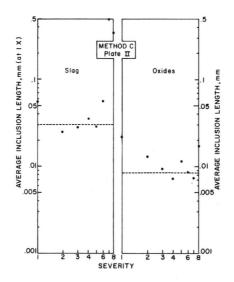

Fig. 33. Average slag and oxide type inclusion length as a function of severity, Plate II, ASTM E45.

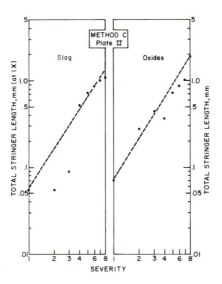

Fig. 34. Total slag and oxide type inclusion stringer length as a function of severity, Plate II, ASTM E45.

Measurement of VDEh 1570–71 Inclusion Chart.  The new German inclusion rating chart, VDEh 1570–71 (Fig. 13), was claimed to be carefully graded, and our measurements confirmed that this was indeed the case.  The chart is divided into four groups of inclusion types, each group having nine severity classifications.  There are two series for sulfides, three for aluminates, three for silicates, and two for globular oxides.  The chart covers steels made in basic-lined furnaces.  The gradations in area and length were established to permit correlation with data from automatic image analysis.  In the two sulfide classes (SS), Column 0 represents elongated sulfides and Column 1 represents sulfides with less elongation and greater width.  For the alumina inclusions (OA), Column 3 shows a single stringer line with considerable elongation. Column 2 illustrates aluminates with a greater amount of elongation and less width, while Column 4 contains a greater number of inclusions per field than Column 3. Three classes of silicates (OS) are also given.  Column 6 is the basic column for measurement with Column 5 illustrating greater elongation and less width and Column 7 containing a greater number of inclusions per field.  For the globular oxides (OG) there are two classes.  Column 8 contains only a single inclusion per field, whereas Column 9 contains a number of inclusions per field.

The severity ratings are designed to cover a wide variety of steels.  The first few rows are applicable to steels of very low inclusion content.  Row 9 describes inclusions of macroscopic size —— 1.5 mm (1/16 inch) or greater.  The area of the inclusions was established so that at any given severity rating the volume fraction of inclusions is equal for Columns 1, 3, 6 and 8.  The area increases according to a power function $(F = 2^n)$, where F is the area and n is the severity rating.  Hence, each time the severity increases, the inclusion area doubles.  Our measurements show that although this doubling is not always precisely achieved, the observed trends are not bad.  The length or diameter measurements have been graded in similar fashion.  With each severity increase the length increases 1–1/2 times. Our measurements show that this uniform increase has not been achieved as planned but that they do increase in a well-graded fashion.

Measurements of number of inclusions per field and per area, total and average inclusion length, and volume fraction are given in Tables 15–18. Figures 35–38 are plots of the area fraction of the inclusions for the four types of inclusions (nine categories). In general, the correlation between volume fraction and severity rating is good. Each curve has some scatter, especially for the 0 severity ratings. Only the ninth column contains considerable nonuniformity. The measurements for the number of inclusions per field and per surface area are shown in Figs. 39–42. There is considerable lack of agreement between the measured data and the severity ratings, but it should be mentioned that this parameter was not intentionally graded. In Figs. 43–46 total and average inclusion lengths are plotted against severity ratings. Total inclusion length correlates reasonably well with severity ratings, but, as would be expected, there is considerable scatter in average inclusion length.

As our data show, although the measured values do not exhibit complete agreement with the stated increments between severity ratings, VDEh 1570–71 is, in general, a well-developed chart method. The inclusion types given relate closely to those observed in steels, and some latitude is provided to handle different degrees of hot working. To sum up, the German chart provides a more meaningful evaluation of inclusions than is possible with the E45 charts.

Efforts to Improve Sensitivity of Chart Methods. Attempts have been made to provide greater sensitivity in JK ratings. For example, Wojcik and Walter[117] used frequency distributions of worst-field JK ratings in such an attempt. A given heat was rated by dividing the number of observations in each severity and inclusion category by the number of fields examined. The shape of the relative frequency distribution curve for each heat was employed to evaluate the inclusion content of each heat and to make comparisons with other heats and heats made by other melting practices. Attempts have also been made to improve the quality of other chart methods, e.g., Firth-Brown Ltd. developed a modified method for using the Fox Chart.

## TABLE 15

### Measurements of elongated sulfides (SS), VDEh 1570–71

| Severity | No./ Field | No/mm² | Total Incl. Length mm | Avg. Incl. Length mm | Volume Fraction % | Severity | No./ Field | No/mm² | Total Incl. Length mm | Avg. Incl. Length mm | Volume Fraction % |
|---|---|---|---|---|---|---|---|---|---|---|---|
| 0.0 | 6 | 12.4 | .115 | .0192 | .13 | 1.0 | 7 | 14.5 | .088 | .0126 | .11 |
| 0.1 | 7 | 14.5 | .288 | .0411 | .37 | 1.1 | 10 | 20.7 | .193 | .0193 | .27 |
| 0.2 | 12 | 24.8 | .473 | .0394 | .51 | 1.2 | 14 | 28.9 | .403 | .0288 | .52 |
| 0.3 | 11 | 22.7 | .491 | .0446 | .63 | 1.3 | 10 | 20.7 | .538 | .0538 | 1.02 |
| 0.4 | 11 | 22.7 | .73 | .0664 | .81 | 1.4 | 9 | 18.6 | .713 | .0791 | 1.45 |
| 0.5 | 16 | 33.1 | 1.208 | .0755 | 1.0 | 1.5 | 17 | 35.1 | 1.011 | .0595 | 1.94 |
| 0.6 | 14 | 28.9 | 1.47 | .105 | 1.37 | 1.6 | 11 | 22.7 | 1.197 | .1088 | 2.58 |
| 0.7 | 13 | 26.9 | 1.528 | .1175 | 1.73 | 1.7 | 15 | 31 | 1.68 | .112 | 2.61 |
| 0.8 | 25 | 51.7 | 2.284 | .0914 | 2.84 | 1.8 | 29 | 59.9 | 2.261 | .078 | 4.37 |

Note: All calculations regarding length or area are at 1X.

## TABLE 16
## Measurements of aluminate inclusions (OA), VDEh 1570-71

| Severity | No./Field | No./mm² | Total Incl. Length (mm) | Avg. Incl. Length (mm) | Volume Fraction % |
|---|---|---|---|---|---|
| 2.0 | 7 | 14.5 | .055 | .008 | .06 |
| 2.1 | 6 | 12.4 | .059 | .01 | .08 |
| 2.2 | 10 | 20.7 | .102 | .0102 | .15 |
| 2.3 | 11 | 22.7 | .174 | .0158 | .28 |
| 2.4 | 17 | 35.1 | .224 | .014 | .45 |
| 2.5 | 25 | 51.7 | .329 | .0132 | .78 |
| 2.6 | 24 | 49. | .389 | .0162 | .77 |
| 2.7 | 35 | 72.3 | .543 | .0155 | 1.07 |
| 2.8 | 39 | 80.6 | .689 | .0177 | 1.59 |

| Severity | No./Field | No./mm² | Total Incl. Length (mm) | Avg. Incl. Length (mm) | Volume Fraction % |
|---|---|---|---|---|---|
| 3.0 | 3 | 6.2 | .034 | .0113 | .07 |
| 3.1 | 7 | 14.5 | .06 | .0086 | .11 |
| 3.2 | 7 | 14.5 | .099 | .0141 | .21 |
| 3.3 | 14 | 28.9 | .186 | .0133 | .30 |
| 3.4 | 14 | 28.9 | .241 | .0172 | .49 |
| 3.5 | 26 | 53.7 | .431 | .0166 | .86 |
| 3.6 | 42 | 86.8 | .669 | .0159 | 1.48 |
| 3.7 | 51 | 105.4 | .95 | .0186 | 2.14 |
| 3.8 | 46 | 95 | .935 | .0203 | 2.67 |

| Severity | No./Field | No./mm² | Total Incl. Length (mm) | Avg. Incl. Length (mm) | Volume Fraction % |
|---|---|---|---|---|---|
| 4.0 | 16 | 33.1 | .116 | .0073 | .10 |
| 4.1 | 26 | 53.7 | .221 | .0085 | .26 |
| 4.2 | 26 | 53.7 | .233 | .009 | .31 |
| 4.3 | 30 | 62 | .348 | .0116 | .62 |
| 4.4 | 42 | 86.8 | .578 | .0138 | 1.03 |
| 4.5 | 36 | 74.4 | .543 | .0151 | 1.1 |
| 4.6 | 52 | 107.4 | .705 | .0136 | 1.35 |
| 4.7 | 69 | 142.6 | 1.09 | .0158 | 2.2 |
| 4.8 | 74 | 152.9 | 1.32 | .0178 | 3.13 |

Note: All calculations involving length or area are at 1X.

## TABLE 17

### Measurement of silicate inclusions (OS), VDEh 1570–71

| Severity | No./Field | No./mm² | Total Incl. Length (mm) | Avg. Incl. Length (mm) | Volume Fraction % |
|---|---|---|---|---|---|
| 5.0 | 1 | 2.1 | .058 | .058 | .06 |
| 5.1 | 1 | 2.1 | .07 | .07 | .07 |
| 5.2 | 1 | 2.1 | .117 | .117 | .12 |
| 5.3 | 2 | 4.1 | .16 | .08 | .25 |
| 5.4 | 2 | 4.1 | .268 | .134 | .47 |
| 5.5 | 4 | 8.3 | .459 | .115 | .79 |
| 5.6 | 4 | 8.3 | .609 | .203 | 1.36 |
| 5.7 | 10 | 20.7 | .913 | .0913 | 2.08 |
| 5.8 | 12 | 24.8 | 1.058 | .0814 | 2.39 |
| 6.0 | 1 | 2.1 | .03 | .03 | .04 |
| 6.1 | 1 | 2.1 | .043 | .043 | .07 |
| 6.2 | 1 | 2.1 | .072 | .072 | .16 |
| 6.3 | 5 | 10.3 | .175 | .035 | .33 |
| 6.4 | 3 | 6.2 | .23 | .077 | .59 |
| 6.5 | 3 | 6.2 | .347 | .116 | 1.1 |
| 6.6 | 3 | 6.2 | .464 | .155 | 1.57 |
| 6.7 | 9 | 18.6 | .952 | .106 | 3.13 |
| 6.8 | 17 | 35.1 | 1.327 | .078 | 4.45 |
| 7.0 | 6 | 12.4 | .055 | .0092 | .09 |
| 7.1 | 13 | 26.9 | .166 | .0128 | .23 |
| 7.2 | 18 | 37.2 | .238 | .0132 | .39 |
| 7.3 | 14 | 28.9 | .338 | .024 | .60 |
| 7.4 | 22 | 45.5 | .524 | .024 | .98 |
| 7.5 | 19 | 39.3 | .816 | .043 | 1.92 |
| 7.6 | 20 | 41.3 | .866 | .043 | 2.34 |
| 7.7 | 23 | 47.5 | 1.123 | .049 | 3.0 |
| 7.8 | 27 | 55.8 | 1.774 | .066 | 4.99 |

Note: All calculations regarding lengths and areas are at 1X.

## TABLE 18
### Measurement of globular oxides (OG), VDEh 1570—71

| Severity | No. Field | No. $mm^2$ | Total Incl. Length mm | Avg. Incl. Length mm | Volume Fraction % | Severity | No. Field | No. $mm^2$ | Total Incl. Length mm | Avg. Incl. Length mm | Volume Fraction % |
|---|---|---|---|---|---|---|---|---|---|---|---|
| 8.0 | 1 | 2.1 | .013 | .013 | .02 | 9.0 | 7 | 14.5 | .033 | .0047 | .085 |
| 8.1 | 1 | 2.1 | .016 | .016 | .032 | 9.1 | 6 | 12.4 | .048 | .008 | .11 |
| 8.2 | 1 | 2.1 | .029 | .029 | .087 | 9.2 | 7 | 14.5 | .052 | .0074 | .15 |
| 8.3 | 1 | 2.1 | .036 | .036 | .20 | 9.3 | 9 | 18.6 | .076 | .0084 | .185 |
| 8.4 | 1 | 2.1 | .06 | .06 | .47 | 9.4 | 9 | 18.6 | .10 | .0111 | .29 |
| 8.5 | 1 | 2.1 | .075 | .075 | .80 | 9.5 | 12 | 24.8 | .132 | .011 | .44 |
| 8.6 | 1 | 2.1 | .092 | .092 | 1.30 | 9.6 | 9 | 18.6 | .222 | .0247 | 1.05 |
| 8.7 | 1 | 2.1 | .129 | .129 | 2.68 | 9.7 | 9 | 18.6 | .238 | .0264 | 1.125 |
| 8.8 | 1 | 2.1 | .19 | .19 | 5.28 | 9.8 | 8 | 16.5 | .307 | .0384 | 2.35 |

Note:  All calculations regarding length and area are at 1X.

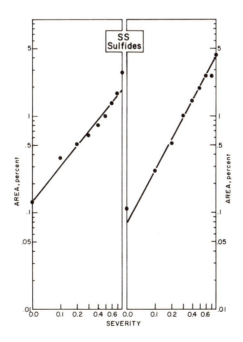

Fig. 35.  Measurement of volume fraction of sulfide (SS) inclusions as a function of severity, VDEh 1570—71.

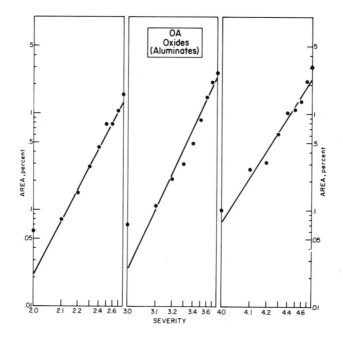

Fig. 36. Measurement of volume fraction of aluminate (OA) inclusions as a function of severity, VDEh 1570–71.

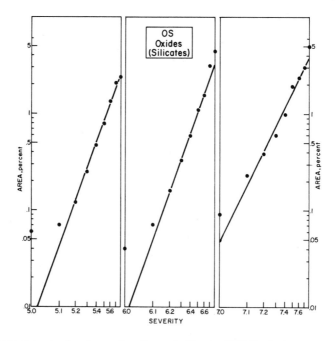

Fig. 37. Measurement of volume fraction of silicate (OS) inclusions as a function of severity, VDEh 1570–71.

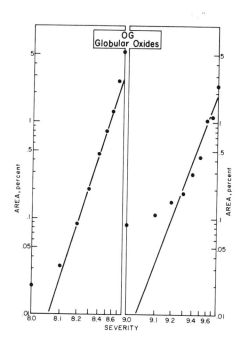

Fig. 38. Measurement of volume fraction of globular (OG) inclusions as a function of severity, VDEh 1570–71.

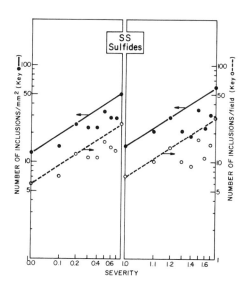

Fig. 39. Number of sulfide inclusions per field and per mm2 as a function of severity, VDEh 1570–71.

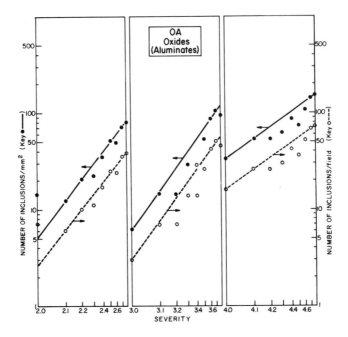

Fig. 40. Number of aluminate inclusions per field and per mm² as a function of severity, VDEh 1570–71.

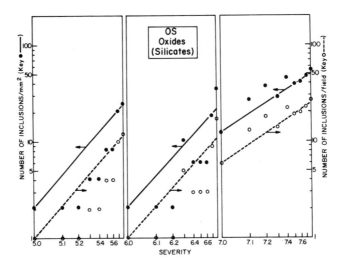

Fig. 41. Number of silicate inclusions per field and per mm² as a function of severity, VDEh 1570–71.

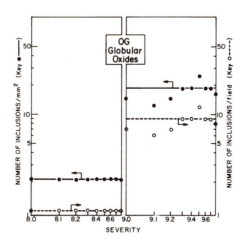

Fig. 42. Number of globular inclusions per field and per mm$^2$ as a function of severity, VDEh 1570—71.

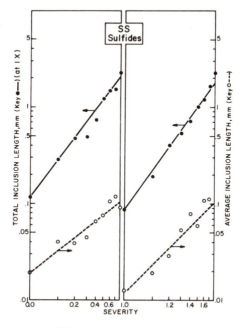

Fig. 43. Total and average sulfide inclusion length as a function of severity, VDEh 1570—71.

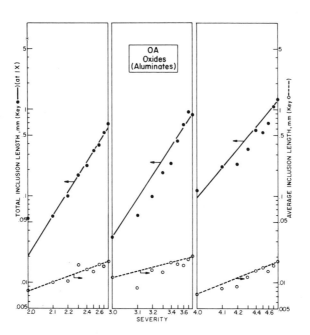

Fig. 44. Total and average aluminate inclusion length as a function of severity, VDEh 1570–71.

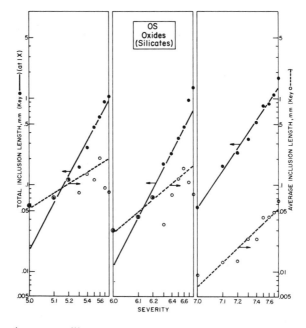

Fig. 45. Total and average silicate inclusion length as a function of severity, VDEh 1570–71.

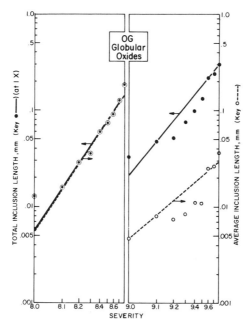

Fig. 46. Total and average globular inclusion length as a function of severity, VDEh 1570–71.

Statistical Evaluations of Chart Ratings. All chart methods suffer from one common difficulty —— limited statistics. Too few samples are examined to provide a realistic measure of inclusion content. Studies[114–116], 118–124] made comparing different rating systems show that accuracy and reproducibility of chart methods are poor. Blank and Allmand[119] compared the JK and Fox Chart methods and two counting procedures. Large errors were observed in the inclusion ratings. The JK ratings were better than the Fox ratings. We measured the pictures on the Fox Chart and found that this chart is indeed inferior to any of the E45 charts. Blank and Allmand's results also revealed considerable disagreement in the classification of oxides by the JK charts.

The study by Mikhalev and Mironov[122] is interesting in that they determined the influence of the number of fields on the measurement statistics. Using GOST 1778, they assessed a large number of samples in a number of heats using an average chart rating and the relative number of field assessments above this average rating. They found that the accuracy of the assessment increases with the number of samples examined from a given heat. For low-carbon steels and ball bearing steels there was an error of ± 10% (with a 90% confidence level) for the average inclusion content obtained with about 50–60 samples per heat. Using rimming heats and the GOST 1778 specification (five levels of severity), they found that it was possible to determine the number of microsections sufficient to separate the worst from the best heats and the average from the worst heat. As expected, it is easier to separate the best from the worst than the worst from the average-quality heat. Of course, from a quality control standpoint it is more important to separate the worst from the average-quality heat. The data (Table 19) show that for 12 samples tested to rate a heat, there is a 45–55% chance that a good heat will be rejected or a bad heat accepted. To distinguish between average- and worst-quality heats about 100 microsamples are required for a reasonably good statistical

TABLE 19

Number of microsamples required to separate rimming steel heats according to inclusion content using GOST 1778 (122)

Probability* of an Equal Assessment, %

| No. of Microsamples | Best Versus Worst Heat | | Average Versus Worst Heat | |
|---|---|---|---|---|
| | Using Avg. Index | Using No Poorer Than Avg. | Using Avg. Index | Using No Poorer Than Avg. |
| 5 | 70 | 85 | 85 | 87 |
| 12 | 45 | 47 | 55 | 55 |
| 20 | 28 | 28 | 40 | 40 |
| 50 | 10 | 10 | 18 | 18 |
| 100 | 2 | 2.5 | 5 | 5 |
| 500 | .1 | .1 | .2 | .2 |

* With a .90 certainty (t=1.68)

separation. These results should be similar for most chart methods. Their conclusions agree with those of Eckstein and Schneider[114], who found that the inclusion assessment based on the Diergarten Chart and a limited number of microsamples did not correlate with other inclusion rating methods.

Use of Different Magnifications with Chart Methods. Hatfield and Giles[123] reported on the influence of magnification on inclusion assessment. A wide variety of magnifications have been used to study inclusions. The JK Chart and Epstein's method[125] were based on 100X. The Fox Chart was based on 130X. The magnification used by Kinzel and Crafts[126] was 50X, Kjerrman's [127] was 200X, and Herty's[128] was 250X. Perhaps the most unusual method from the standpoint of choice of magnification was Wilandh's "picture-drawing" method[129] with three magnifications (67,260 and 670X) chosen to conform to a ratio of 1; $\sqrt{10}$:10 in order to produce area ratios of 100:10:1. Other studies, such as that of Bergh and Lindberg[130], utilized observation at 100 and 500X. In their evaluation of the influence of magnification, Hatfield and Giles[123] made inclusion measurements with the Fox Chart at 27.5, 132, 520, 1200 and 2000X. Counts were made unetched and after a very light etch with dilute picral (10 sec. in 30% of the normal 4% solution). They observed marked increases in the inclusion counts on some samples after etching and a consistent decrease in the inclusion count as the magnification increased. They attributed this trend in part to the poor geometrical basis of the chart.

Manual Inclusion Rating Methods. The problems inherent in chart ratings of inclusion content have fostered the development of manual methods. Although these methods are based on quantitative metallographic techniques, rigorous adherence to the theoretical precepts of stereology have not always been followed. Consequently, some of the methods produce results that are qualitative. In this review we will describe only those methods that are based on a firm stereological basis.

They are categorized as:
- Manual measurement of inclusion content
- Size measurements
- Number of inclusions per volume
- Statistics of inclusion size and distribution
- Sizing of inclusions using transmission electron microscope

Manual Measurement of Inclusion Content. Many researchers have developed their own procedures for assessing inclusion content. Methods have been proposed by Epstein[125], Kinzel and Crafts[126], Kjerrman[127], Herty *et al*[128], Zieler [131], Jolivet[132], Hunter[133], and Allmand[91]. Allmand reviewed these methods as well as the Fox, JK, Diergarten and Walker chart methods. Other methods have been developed by Whiteley[134], Swinden[135], Wilandh[129], Bergh and Lindberg[130], Maurer and Schustek[136], Baynes[137,138], Welchner and Hildorf [139], Aronovich and Lyubarskii[140], Czochralski[141] and Hardy and Allsop[118].

Epstein's method[125] has been incorporated into ASTM E45 as Method B. This method was adopted in 1942 by the Society of Automotive Engineers (SAE) [142–143] but has been replaced by the Walker Chart method (plate II of E45). The counting method proposed by Hardy and Allsop[118] was modified by the BISRA Nonmetallic Inclusion Group[144] and is referred to as the lineal-traverse method. This method has been widely used and compares favorably with other methods[cf. 145 and 119].

Of the many standards reviewed in this paper, only the Japanese Specification JIS G 0555 incorporates a method of assessing inclusion content based solely on stereological principles[103]. This specification calls for a point counting technique similar to that of Bergh[141] and that of Bergh and Lindberg[130]. In the Japanese method the inclusions are divided into three categories:

A—Type:  Plastically deformed inclusions can be subdivided as $A_1$ (sulfide) and $A_2$ (silicate)
B—Type:  Discontinuously elongated inclusions
C—Type:  Randomly scattered undeformed (i.e., globular) inclusions.

An eyepiece reticle with twenty horizontal and vertical lines is used. Field selection is done randomly, with at least 30 and preferably 60 fields. The number of grid points occupied by inclusions is counted. A magnification of 400X is recommended. The percentage of area occupied by the inclusions is determined by the following equation:

$$d = \frac{n}{p \times f} \times 100$$

where d is the index of cleanliness in %
    n is the number of grid points occupied by the inclusions
    p is the total number of grid points on the reticle, and
    f is the number of fields.

The results are expressed as the index for all inclusions and the index for each type.

The method proposed by Bergh[146] is somewhat similar although more complex. An eyepiece reticle with a grid having a very large number of intersections is used. For each field the number of grid intersections occupied by inclusions is counted and each inclusion is measured in terms of the grid dimensions. The grid is positioned at a 30° angle to the hot-working axis. From these measurements the percent volume and the number of inclusions per unit area as a function of the inclusion size can be calculated. The method also permits calculation of the mean length and width of the inclusions, the aspect ratio, and an equivalent diameter (making assumptions as to the shape of the inclusions).

Classification schemes for particular types of steel have been proposed. Kammori et al[147] suggested a method applicable to rimming steel ingots. Methods for classifying inclusions in castings were proposed by Baynes[137], Buzek and Schindlerova[148], and Morykwas[149].

Size Measurements. In the manufacture of free-machining steels the shape of the manganese sulfides is an important characteristic, as shown by the numerous studies of inclusion malleability. In a study of the processing of AISI B1113 free-machining steel, Carney and Rudolphy[150] used a procedure similar to Van Vlack's [151] to measure the length-width ratios and the number of inclusions in the ingot, bloom and billet samples. Inclusions greater than 0.007 mm in length were counted, and the relative size of inclusions was calculated by means of the length and width measurements and the equation $V=1.08\ LW^2$ for the relative inclusion volume. At least 50 inclusions were measured. The relative size of the average inclusion was then calculated.

Number of Inclusions per Volume. Bandyopadhyay et al [152] studied the average size and quantity of inclusions during the course of melting. Since the samples were as-cast, the simplifying assumptions regarding shape are closer to reality than in the case of a wrought steel sample. The size of the inclusions was measured using an eyepiece graticule. At least 40 randomly selected inclusions were measured on each sample. The average number of inclusions per unit area was determined from direct counting of at least ten fields per sample. The number of inclusions per unit volume, $N_v$, was calculated from an equation and the measurements. Inclusion-area fraction was determined by point counting. In the derivation of the $N_v$ equation all inclusions were assumed to be spherical with a radius r. The measured average inclusion size (diameter), $\bar{D}_s$, was shown to be related to the true radius of the spherical inclusions by:

$$\bar{D}_s = \pi r/2$$

The number of inclusions per unit surface area, $N_A$, is related to the number of inclusions per unit volume by:

$$N_A = 2rN_v$$

The area fraction of inclusions, $A_A$, is related to the number of inclusions per unit area and their true diameter, $D_s$, as follows:

$$A_A = \frac{\pi D_s^2 N_A}{4}$$

The true number of inclusions per unit volume can be calculated from the following equation using the measured average diameter, $\bar{D}_s$, and the area fraction, $A_A$:

$$N_v = \frac{N_A}{2r} = \frac{4A_A/\pi\ \bar{D}_s^{\ 2}}{4\bar{D}_s/\pi} = \frac{A_A}{\bar{D}_s^{\ 3}}$$

Statistics of Inclusion Size and Distribution. The statistical treatment of inclusion ratings must cope with the following:
- Nature of distribution of inclusions
- Amount of sampling

- Location of samples
- Surface area examined
- Specimen preparation
- Measurement method
- Operator ability.

With the exception of the first of these, all of the above variables have already been discussed here. The counting of the number of inclusions per field of view has been shown[137,153,154] to follow a Poisson-type distribution, i.e., one that describes the occurrence of isolated events in a continuum.

Pelazzi[153] performed inclusion counts on as-cast 0.50% carbon steel. He observed that the counts of the number of inclusions per field exhibited a Poisson distribution, as verified by the chi-squared test. This distribution was also observed if the inclusion counts per field were additionally categorized by inclusion types or by size. Such measurements were performed at both 100 and 700X magnification. He showed that for steels that contain a high number of inclusions per field, as in highly resulfurized steels, the distribution approaches a normal distribution. Pelazzi developed an equation to determine the number of fields, N, to be examined in order to obtain a desired coefficient of variation:

$$N = \frac{t^2 [CV]^2}{V_u^2}$$

where t is a tabulated value for a normal distribution (t=1.96 for a 95% confidence level and t=1.65 for a 90% confidence)
CV is the coefficient of variation: $100/\sqrt{u}$ (u=mean) and
$V_u$ is the desired relative error.

As an example of the use of the equation, suppose all of the inclusions in 20 fields have been counted, yielding a mean of 9 inclusions per field. The coefficient of variation is $100/u^{1/2}$ or 33.3%. To find the number of fields required to give a 10% relative error with a 95% confidence, the following calculation is made:

$$N = \frac{(1.96)^2 (0.333)^2}{(0.1)^2} = 43 \text{ fields}$$

To achieve a 5% or a 2% relative error, one would be required to count all of the inclusions on 170 or 1065 fields, respectively. One can use the equation to calculate the relative error of a given measurement by rearranging the formula and solving for $V_u$:

$$V_u = \frac{t^2 [CV]^2}{N}$$

If one develops size-frequency histograms as done by Bergh[1] and by Simpson and Standish[155,156], they do not resemble a normal distribution. Instead, the distribution by size is highly skewed toward the smaller sizes, thus exhibiting a lognormal distribution. The right-hand tail of this distribution, representing the number of large inclusions, is of greatest interest in structure-property correlations since they generally exert the strongest influence on properties. Simpson and Standish also examined the spatial distribution of inclusions and found that it also follows a lognormal distribution. They observed that the distribution of clustered aluminates is quite different from that of randomly dispersed aluminates with respect to the mean value and variance.

Sizing of Inclusions Using Transmission Electron Microscopy. A number of researchers[157–162] have performed quantitative measurements of inclusions on ductile fracture faces, and several investigators[157,161,162], using replicas examined on the transmission electron microscope, have correlated these measurements to the size and spatial distributions of the inclusions measured on metallographic specimens. In general, these studies concern inclusions in weld metal that are spherical and randomly distributed — ideal specimens for such studies. Steel[163] developed an electropolishing method applicable to polished steel weld metal that produces a flat surface and leaves the inclusions unattacked and in relief. In this method, the specimen is ground through the standard range of papers and electropolished in a 6% perchloric acid solution (2000 ml methanol, 25 ml distilled water, 150 ml perchloric acid and 60 ml ether) with a potential of 43 volts and a current density of 0.01 amperes/sq. mm for 30 seconds. Next, the sample is etched in 2% nital for two seconds to produce some relief in the surface of the matrix. The samples are washed with distilled water and ethanol. Extraction replicas are made using a single-stage carbon-film technique. With care, this method is capable of a very high extraction efficiency. Counting and sizing are conducted on TEM photographs, generally manually. The volume fraction of inclusions determined by this process agrees well with measurements made by more traditional methods. Widgery and Knott[157] present sizing distributions from: a polished sample electropolished and replicated, a replicated fracture, and a bulk sample. The agreement among the three methods is quite good.

Inclusion Measurement Using Standard Stereological Methods

The standard procedures used in manual quantitative metallography (stereology) may be used to estimate inclusion content, number, distribution and size. These methods have been described in numerous papers and text books on the subject* and are categorized as follows:
- Volume fraction measurement
- Number of inclusions per surface area
- Distribution of inclusions
- Number of inclusions per volume.

Volume Fraction Measurement. The inclusion volume fraction can be estimated using point counting or lineal analysis procedures. Data utilizing these procedures are given in the following section in comparison with similar data developed from image analysis. In the point count method a grid with a number of systematically located points is placed over the projected image or on a photograph. The number of points lying on inclusions is counted. This procedure is repeated for a number of fields. The total number of points lying on inclusions is divided by the number of grid points times the number of fields to produce an estimate of the volume fraction of inclusions. This estimate is generally expressed as a percentage by multiplying by 100. Grid points tangent to inclusions are counted as ½. This method was adopted by the Japanese in their specification JIS G 0555.

The lineal analysis method uses a grid overlay with a number of uniformly spaced parallel straight lines of known length. The length of these lines lying within the inclusion is divided by the total line length to estimate the volume fraction of inclusions. The method generally employs a device such as a Hurlbut counter, the use of which facilitates the measurement. For low volume fraction measurements the lineal analysis method is much more tedious to use than point counting.

---

* See *Quantitative Stereology* by E.E. Underwood, Addison-Wesley Publishing Co., Reading, Massachusetts, 1970.

Number of Inclusions Per Surface Area. The number of inclusions within a known area can be counted to give an estimate of $N_A$ (number per unit area). As previously mentioned, this measurement varies according to the sectioning plane orientation. The volume fraction measurements, however, are not influenced by the plane-of-polish orientation, but the calculated accuracy is. For a given number of field measurements the accuracy of volume fraction measurements is better on transverse than on longitudinal planes. Methods for calculating statistical accuracy are given in the next section.

Distribution of Inclusions. A useful technique for measuring the distribution of inclusions is provided by the mean free path $\lambda$ (mean edge-to-edge distance between particles). To estimate $\lambda$, two measurements are made — the volume fraction and $N_L$, the number of particles intersected per unit length of test line. The volume fraction can be estimated using point counting or lineal analysis. To estimate $N_L$ a grid overlay identical to that for lineal analysis is placed over the projected image. The number of inclusions intersecting the lines is counted for n fields. The number of inclusion intersections is divided by the total true line length to give $N_L$. Next, the mean free path is calculated as follows:

$$\lambda = \frac{1 - V_v}{N_L}$$

Number of Inclusions Per Volume. Measurements of the number of inclusions per unit volume of material, $N_V$, can be made by manual stereological methods. These procedures are complex and time-consuming. Their accuracy is limited due to assumptions that must be made regarding inclusion shape. They are simplest to perform on as-cast samples, where the inclusion shapes generally conform better to simple shapes such as spheres. The study by Bandyopadhyay et al [152] is one example of the application of such procedures.

Automatic Image Analysis

The tedious nature of these manual methods and their limited statistical value have generally militated against their use in quality control and research studies. As the cited references show, a great deal of effort has been expended on trying to develop fast reliable methods for inclusion assessment. By far the most promising approach has resulted from the development of automatic image analyzers [164–173], which have been used in many inclusion studies [113,174–212] These devices greatly reduce the subjectiveness of inclusion ratings, while their speed permits obtaining acceptable statistical data within a reasonable time. Unfortunately, these devices are quite expensive and require skilled operators and careful attention. The Quantimet 360 has been designed as a special-purpose instrument for grain-size and inclusion rating and is the simplest of these instruments to operate. Nearly all of the other instruments are flexible multipurpose instruments, which are more complex and slightly slower in operation [209]. At present, methods utilizing image analysers in specifications have not been standardized. However, work is under way to do so.

Our review of the use of image analysis for inclusion measurement is covered in the following categories:
- Influence of the nature of the intended study
- Inclusion detection
- Choice of magnification
- Number of fields measured
- Correlations with chemical analysis
- Size measurements
- Measurement of number per volume

- Combined image analysis and composition measurement
- Statistical analysis of data
- Example of inclusion measurement and data analysis.

Influence of Nature of Intended Study.  The application of image analyzers for quality control will be different than for research studies.  In quality control studies one generally wants to know whether a given heat or lot is within certain limits regarding inclusion content.  Such decisions may be based on statistics describing the inclusion content in terms of a mean and its standard deviation.  Features of interest would include the volume fraction (total and by type of inclusion) and stringer lengths greater than some critical size.  In quality control studies the melting, teeming and hot-working practices for a given product are usually fixed, a fact that establishes certain constant features.  Hence, the quality control study must be geared to detect deviations from the norm.  In research studies the processing parameters may be varied hence the analytical techniques must be sufficiently general to describe the inclusion population and document changes due to practice modifications.

Inclusion Detection.  The detection of inclusions is based on their reflectivity and is influenced somewhat by the sample preparation.  Franklin[199] summarized reflectivity data for inclusion types that may be found in steels and showed them in reference to an as-polished steel surface (Fig. 47).  The reflectivity of a hardened steel surface is generally reported as 59%.  Since some of these inclusions are partly transparent, reflections may come from the underlying steel matrix, thus creating a bright spot in the center of the inclusion.  This is a particular problem in the detection of large, glassy silicates.  Some of the latest image analyzers can overcome this problem by use of a software-generated "hole-closing" program.  In this manner, bright spots within a glossy silicate inclusion can be detected.

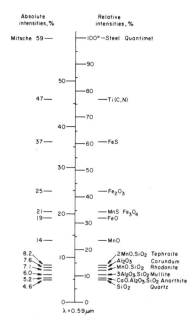

Fig. 47.  Reflectivity of inclusions commonly found in steels[199].

As the data show, the reflectivity of manganese sulfides (21%) is considerably higher than that of most of the oxides usually present. The range of reflectivity of the oxides generally present in steel is about 4.6–14%. Fortunately, for a given deoxidation practice, one usually has to separate manganese sulfides from only one oxide type.

Relief around inclusions will increase the apparent area of the inclusions, thus leading to higher volume fractions. Johansson [192] compared the reproducibility of inclusion measurements on samples polished manually or automatically and found that the manually polished samples exhibited higher inclusion content due to relief. Automatic polishing devices with low nap cloths are recommended. In our experience, analysis can be conducted after a 1 $\mu$m diamond final polish. Going to finer polishing compounds, such as 0.3 $\mu$m alumina, is not necessary because the scratches produced by 1 $\mu$m diamond, especially if the sample is hardened, do not affect the inclusion measurement. These finer polishing solutions often introduce relief and should be avoided whenever possible.

In many cases one is able to separate inclusion types by their reflectivity. This becomes more difficult as the inclusion size decreases. Some investigators have used selective etchants (Table 20) to help separate the inclusion types. In particular, sulfides have been discriminated using dilute (1.5 or 10%) aqueous chromic acid [108,213].

Another approach would be to measure all inclusions and then preferentially attack the sulfides with a 5% $H_2SO_4$ etch [108] that dissolves the sulfides. Then the remaining inclusions would be remeasured. Beraha's staining reagents [214,215] could be used to selectively stain sulfides for measurement. This etch procedure has been used by some investigators [216–218] but has not yet been documented in image analysis studies.

Whiteley [27] developed a method to distinguish manganese sulfide from ferrous oxide. Both inclusions are similar in color, with FeO being slightly darker and slightly less reflective(Fig. 47). An unused selvyt cloth is soaked for several minutes in a freshly prepared 5% aqueous solution of silver nitrate. The cloth is then thoroughly washed under running water to remove all unadsorbed silver. The cloth is placed nap side up on a glass plate and the polished sample is gently rubbed on the cloth. Generally, 100–150 strokes in one direction or a quick to-and-fro motion for 15 seconds is adequate. Some of the adsorbed silver will be transferred from the cloth to the MnS (or FeS) inclusions. Under bright-field illumination the sulfides look white. The impregnated cloth is good for a number of samples and can be used for several days.

The method of Gerds and Melton could probably be used to preferentially distinguish lead inclusions, but polarized light must be used. Chalfant [43] showed that lead inclusions appear off-white after final polishing with a weak solution of gamma-alumina adjusted to a pH of 7. Volk [220] developed an etchant (Table 20) that colors lead inclusions yellow by the formation of lead iodide. Bright-field illumination can be used, but polarized light produces better results. An electrolytic etching technique [221a] was developed by Bardgett and Lismer to reveal the presence of lead inclusions (Table 20). Schofield [221b] also developed an etchant for revealing lead inclusions (Table 20). Sarracino and Rossi [42] reviewed a number of techniques for identifying lead inclusions (Table 20).

Another approach for inclusion detection is the deposition of a highly refracting dielectric compound such as ZnSe or TeSe to produce interference films. This method, developed by Pepperhoff [222,223] has been used by Johansson [191], Allmand

## TABLE 20
### Etchants used to distinguish certain inclusions

| | Etchant | Use | Reference |
|---|---|---|---|
| 1) | 15g $CrO_3$<br>1000 cc $H_2O$<br>Approx. 60 sec.<br>(10% $CrO_3$ solution also used) | Sulphides | 108, 213 |
| 2) | 5% $H_2SO_4$ in $H_2O$ | Dissolves sulphides | 108 |
| 3) | 240g sodium thiosulfate<br>24g lead acetate<br>30g citric acid<br>1000 ml $H_2O$<br>Add 200 mg sodium nitrite per 100 ml when ready to use, good for 1/2 hr.<br>(Allow to stand 24 hr. before use, store in dark bottle.) | Sulphides are<br>silvery bright | 214, 215 |
| 4) | Etch first with 4% Picral<br>Soln. A:  1g potassium cyanide<br>          100 ml $H_2O$<br>Soln. B:  0.25g dithizone<br>          10 ml chloroform<br>(Mix A with B, swap etch up to 3 min.)<br>Pb inclusions are red in polarized light | Lead | 218 |
| 5) | 50 ml alcohol<br>3 ml glacial acetic acid<br>20 ml water<br>2g potassium iodide<br>(immerse 1-2 min.) | Colors Pb yellow | 220 |
| 6) | 10% aqueous ammonium acetate solution<br>electrolytic, 30 sec. @ 5V | Reveals Pb | 221a, 42 |
| 7) | 1-2 ml Nitric Acid<br>98-99 ml alcohol<br>saturate with potassium iodied,<br>filter off excess KI<br>immerse in filtrate 10-30 sec. | Pb<br>Greenish yellow in BF<br>Bright yellow in DF | 221b, 42 |
| 8) | 10% chromic acid in water<br>immerse 10 min. | · Colors Pb yellow red | 42 |
| 9) | Immerse sample one hour in mercury<br>Heated to 100 C | Attacks Pb | 42 |

and Houseman[38], Buhler *et al*[224], and Silber and Judtmann[39]. A small lump of the substance chosen is evaporated under vacuum with a coating device for preparing samples for electron microscopy. The present writer has used this method and found it to be highly useful. For inclusion studies no pre-etching or relief is required. Careful ultrasonic cleaning is recommended to give a uniform deposition. Slowly evaporate the substance onto the surface of the specimen while observing the color change (yellow-red-blue). It is best to stop the evaporation when a violet or purple color is obtained, since this gives the best color contrast. This procedure produces a layer thickness of about 30–50 nM. The evaporated layer is highly refractive. Interference also occurs at the layer/metal interface. The net result is a marked contrast between the differently colored individual phases.

Choice of Magnification. Once a detection scheme has been established, the operator must decide on the magnification to use and the number of fields to measure. The selection of the magnification is a compromise between the surface area that can be examined in a given time and the resolution and ease of detection. Low magnifications permit large areas to be examined in a short time. This advantage is helpful in minimizing statistical error, because the field-to-field measure of volume fraction or inclusion number is more consistent at low magnifications. However, since inclusions are generally small, very low magnifications may not provide enough resolution. Also, a small variation in threshold setting (gray level) produces a larger measurement error with low magnifications than with higher magnifications. At higher magnifications it is easier to set the threshold and also to discriminate inclusion types. However, each time the magnification is doubled, the field area is reduced to one-fourth its original size. Hence, to cover the same area, four times as many fields must be examined.

Allmand and Coleman[201] examined the statistical variation in samples cut from a steel billet using a 5X objective. Their data showed a high amount of scatter. In practice, the magnification at the television screen is not simply the product of the magnification of the objective and projection eyepiece, because there is also a magnifying factor in the television camera. Roche[205] used the same image analyzer but showed the statistical deviation for a number of objectives versus the number of fields measured, and his data exhibited better accuracy. Allmand and Coleman concluded that at least 400 fields on each sample are required to obtain a reasonable accuracy. Roche's data for 10, 20 and 40X objectives and inclusion volume fractions up to about 1% indicate that substantially fewer fields are required to obtain a 10% relative accuracy with a 95% confidence. Part of this difference between the results of these two studies may be due to differences between the samples investigated in each study. Our own data tend to agree more closely with Allmand and Coleman's observations.

Number of Fields Measured. Rege *et al*[190] published results showing the effect of the number of fields tested on measurement accuracy. In their study they first tested the entire sample surface with 1000 fields (10 across by 100 in the rolling direction). They next surveyed the same sample using 50 fields (5 widely spaced locations by 10 across), 90 fields (9 locations by 10 across), and 190 fields (19 locations by 10 across). In each measurement, the volume fraction of inclusions and the volume fraction of the worst field were recorded. The estimate of inclusion volume fractions with 50, 90 or 190 fields gave good correlations with the 1000-field estimate. However, the volume fraction measurement of the worst field using 50, 90 or 190 fields was lower than the worst-field volume fraction measurement using 1000 fields. They found that the volume fraction of the worst field using 50, 90 or 190 fields increased as the number of fields increased. These authors also presented a method for assessing the statistical accuracy of the test data.

Hersant and Jeulin[203] developed a unique technique for determining the accuracy of inclusion measurements and for estimating the number of fields required to obtain a certain level of accuracy. The method is based on the geostatistical theories of G. Matherton and utilizes point variograms determined in two directions, parallel and perpendicular to the rolling direction, and ordered variograms based on the field sampling scheme. Results are given for sulfide measurements in five samples of carbon steel (silicon-killed) as shown in Table 21. These results show that a relatively high number of fields must be examined to achieve good statistical accuracy.

TABLE 21

Number of fields required to assess sulfide content[203]

| Sample No. | No. of Fields for Given Accuracy (95% Confidence) | | |
|---|---|---|---|
| | ±5% | ±10% | ±20% |
| 1 | 8370 | 2099 | 523 |
| 2 | 3174 | 794 | 199 |
| 3 | 4448 | 1112 | 278 |
| 4 | 2578 | 645 | 162 |
| 5 | 3504 | 876 | 219 |

Correlations with Chemical Analysis. Numerous image analysis studies have shown that calculation of the weight percent sulfur or oxygen in the inclusions based on image analysis measurements agrees well with the sulfur and oxygen content determined by traditional analytical techniques. Referring to the work of Roche[175] as an example, the weight percent of oxygen is calculated as follows. (Weight percent sulfur is similarly calculated). First, after a measurement of the volume fraction of the oxides, $(V_V)$ oxides, the volume fraction is converted to a weight fraction:

$$(Wt\ \%)_{oxides} = \frac{d_{inclusion}}{d_{matrix}} (V_V)_{oxides}$$

where $d_{inclusion}$ is the density of the inclusion, and

$d_{matrix}$ is the density of the metallic matrix

(the density of iron is 7.87 g/cu cm).

Next, the proportion by weight of the oxygen in the inclusions is calculated:

$$Wt\ \%\ Oxygen = \frac{M_{O_2}}{M_{inclusion}} (Wt\ \%)_{oxides}$$

where $M_{O_2}$ is the weight of oxygen in a mole of inclusion, and
$M_{inclusion}$ is the molar weight of the inclusion.

Since all of the above factors, except for the volume fraction, are constant for a given inclusion type, these factors can be combined into a single constant, which Roche referred to as k. Then, the weight percent oxygen is calculated by multiplying the volume fraction of the oxides by the particular k value. Roche calculated k for a number of inclusions (Table 22).

## TABLE 22
### K—values for common inclusion types[175]

| Inclusion Formula | Name | Density $(g/cm^3)$ | k |
|---|---|---|---|
| FeO | Wustite | 4.9 - 5.3 | 0.148 |
| MnO | Manganosite | 5 | 0.144 |
| $Al_2O_3$ | Corundum | 3.9 - 4.1 | 0.235 |
| $FeO \cdot Al_2O_3$ | Hercynite | 3.91 - 3.95 | 0.188 |
| $3Al_2O_3 \cdot 2SiO_2$ | Mullite | 3.03 | 0.190 |
| $SiO_2$ | Quartz / Tridymite / Cristobalite | 2.27 - 2.65 | 0.165 |
| $2\,FeO \cdot SiO_2$ | Fayalite | 3.91 - 4.34 | 0.167 |
| $2MnO \cdot SiO_2$ | Tephroite | 4.04 | 0.163 |
| $CaO \cdot P_2O_5$ | -- | 2.82 | 0.160 |
| $3CaO \cdot P_2O_5$ | -- | 3.14 | 0.259 |
| $(FeO, MnO)\,Cr_2O_3$ | -- | 4.5 - 4.8 | 0.170 |
| $Cr_2O_3$ | -- | 5.21 | 0.210 |

In like manner the weight percent of sulfur in the sulfide inclusions can be calculated with the density of manganese sulfide as 4.0 g/cu cm. The weight percent sulfur in the manganese sulfides is calculated as:

$$Wt\ \%\ Sulfur\ =\ \frac{32.066}{87.006}\ \times\ \frac{4.0}{7.87} \times\ (V_v)_{sulfide}\ =\ k(V_v)_{sulfides}$$

In this case the k for MnS is 0.187. Since there is a small amount of sulfur dissolved in the iron, the weight percent of sulfur in the sulfide inclusions is slightly less than the total sulfur content.

Size Measurements. Many investigators have performed size measurements of inclusions. As previously mentioned, the size of inclusions in wrought steels varies with the orientation of the test plane. Hence, studies have been done with either as-cast steels, where the inclusions are basically spherical, or on longitudinal sections from wrought steels. With inclusions in wrought steels the mean inclusion length and width have usually been measured to develop a mean shape factor, or aspect ratio. Some studies have developed histograms of inclusion lengths.

Calculation of a mean aspect ratio based on field averages of inclusion length and width is not as good a procedure as measuring the length and width of each inclusion to calculate the aspect ratio of each inclusion and then obtain a mean aspect ratio based on individual measurements of a number of inclusions. Some of the current image analyzers are capable of making individual measurements.

An interesting approach to the measurement of sulfide inclusions in free-machining steels was Frohlke's[197]. This author was not satisfied with attempts to relate mean inclusion length and breadth or mean aspect ratio to machinability. Instead, Frohlke used transverse specimens and measured the diameter of the sulfides. He found that fewer particles needed to be measured to develop a size distribution for transverse specimens than for longitudinal samples. The sulfides were graded into eight size classes based on their diameter. The number of sulfides in each size group was determined for 45 fields using a 32X objective and a 5X ocular. Based on the total measurement area, the mean number of particles per sq mm was calculated for each size group. Frohlke found a straight-line relationship between the class limits of the sulfide diameters, $d_i$, and the number of sulfides per sq mm, $N_i$:

$$\log N_i = -md_i + b$$

This straight line was referred to as a "manganese sulfide identification line" (MIL). He used his technique to investigate the effect of sulfide composition and amount of deformation on sulfide size distribution.

Measurement of stringer lengths or cluster size is rather easy to do manually but difficult to do with image analysis. With some of the modern image analyzers it is possible to dilate and reconstruct the structure so as to fuse neighboring inclusions. Furthermore, the amount of dilation can be easily and reproducibly controlled. Using such procedures, we were able to correlate inclusion cluster size with the size of ultrasonic indications.

Measurement of Number per Volume. Few studies have been conducted to determine the true number of inclusions per unit volume, $N_V$. Due to the shape assumptions that must be made, most such studies have utilized as-cast samples. One obvious way to determine $N_V$ is by serial sectioning. However, we are unaware of any such studies that have been published. The standard stereological methods that have been developed for spherical shapes may be applied to determine $N_V$ for spherical inclusions. In determining $N_V$ for ellipsoids, the methods of Wicksell[225,226] and DeHoff[227] could be applied. A complete description of these methods is beyond the scope of this paper. Such measurements, however, are among the most difficult to make, because they require considerable time when done manually. Once the necessary programming has been accomplished, the use of image analyzers with minicomputers greatly reduces the effort required. DeHoff[228] developed a somewhat simpler method for determining $N_V$ and the mean geometric diameter and standard deviation of particles with the same similar shape. His method requires three measurements: the volume fraction of particles ($V_V$), the number of particles per unit area ($N_A$), and the number of point intersections with the particles per unit length ($P_L$). If the particles measured conform closely to the assumed shape, good accuracy can be obtained.

Combined Image Analysis and Composition Measurement. A truly unique approach to the problem of inclusion identification and measurement has been provided by the PASEM (Particle Analyzing Scanning Electron Microscope) developed in Sweden[229,230]. In this system a minicomputer controls a scanning electron microscope, an energy-dispersive x-ray spectrometer, a special-purpose digital image memory, and a disc-storage module. Instead of using an optical microscope to form an image that is analyzed according to inclusion reflectivity, the SEM image is formed using backscattered electrons. Image contrast is due to differences in atomic number. The SEM image is stored in memory along with the coordinates, area, perimeter, and x and y Feret diameters for detected objects. The large depth of focus of the SEM is particularly helpful in eliminating focusing problems at high magnifications. The size measurements can be combined with the analytical data to provide a complete description of the inclusions. Particles down to ~0.2 $\mu$m are analyzed. The combination of chemical analysis and image analysis does require considerable time —— analysis of several hundred particles requires a few hours (5—10 seconds per inclusion).

Statistical Analysis of Data.  An important aspect of image analysis is the calculation of the accuracy of the measurements.  The procedure utilized is that recommended by DeHoff[231,232].  The analysis is begun by calculating the mean, $\bar{X}$, and the standard deviation, $\sigma$, of the individual field measurements (volume fraction, number per field or per area, etc.):

$$\bar{X} = \frac{1}{n} \sum_{i=1}^{n} x_i$$

where $x_i$ represent the values of the individual measurements, and
        n is the number of fields.
Next, the standard deviation of the observations is calculated:

$$\sigma = \left[ \frac{1}{n-1} \sum_{i=1}^{n} (x_i - \bar{X})^2 \right]^{\frac{1}{2}}$$

The value $\sigma^2$ is referred to as the variance.  A simplified form for calculating $\sigma^2$, which is more amenable to pocket calculators, is:

$$\sigma^2 = \frac{1}{n(n-1)} \left[ n \sum_{i=1}^{n} X_i^2 - \left( \sum_{i=1}^{n} X_i \right)^2 \right]$$

The standard deviation of the mean; $\sigma\bar{X}$ is then calculated:

$$\sigma\bar{X} = \frac{\sigma}{\sqrt{n-1}}$$

This relationship indicates that the standard deviation relative to the mean is inversely proportional to the square root of the number of measurements.  That is, in order to reduce the standard deviation of the mean by one-half, four times as many measurements are required.  To determine the 95% confidence limit, multiply $\sigma\bar{X}$ by 1.96 (for 90 or 99% confidence limits, the multiplying factors are 1.645 and 1.576, respectively).  The results can be expressed as the mean value plus or minus the desired confidence limit, as for example:

$$\bar{X} \pm CL_{95}$$

The relative accuracy in percent can be expressed by dividing the desired confidence limit by the mean value and multiplying by 100:

$$\text{Percent relative accuracy} = \frac{CL_{95}}{\bar{X}} \times 100.$$

DeHoff[231] suggested a simple method for determining the approximate number of measurements required to obtain a desired relative accuracy once an estimate of the mean and standard deviation has been made:

$$n = \left[ \frac{200}{\%\text{Rel. Acc.}} \cdot \frac{\sigma}{\bar{X}} \right]^2$$

In the ASTM development of a standard method for inclusion assessment based on image analysis, G.A. Moore has criticized this procedure, because the distribution of the volume fraction measurements about the mean does not conform to a normal distribution.  To eliminate this problem, Moore suggests grouping the field measurements into a number of sets.  Each set would consist of from 30 to

100 sequential field measurements. A set should consist of one or more rows of contiguous fields placed across the specimen perpendicular to the hot-working direction. From 8 to 12 such sets should be measured. The set averages are used to calculate the statistics using the four-moment method, which provides a measure of the normalcy of the data.

The four moments are calculated from the set averages ($X_i$, $X_2$, ... $X_n$), as follows:

$$A_1 = \Sigma (X_1 + X_2 + X_3 + ... + X_n)$$

$$A_2 = \Sigma (X_1^2 + X_2^2 + X_3^2 + ... + X_n^2)$$

$$A_3 = \Sigma (X_1^3 + X_2^3 + X_3^3 + ... + X_n^3)$$

$$A_4 = \Sigma (X_1^4 + X_2^4 + X_3^4 + ... + X_n^4)$$

where $A_1$, $A_2$, $A_3$ and $A_4$ are the four moment sums.

The four set sums are then each divided by the number of sets to produce the four moments about zero:

$$V_1 = A_1/n$$

$$V_2 = A_2/n$$

$$V_3 = A_3/n$$

$$V_4 = A_4/n$$

$V_1$ is the arithmetic mean of the set measurements.

The RMS average, which represents the influence of inclusion segregation, is calculated as:

$$\tilde{X} = (V_2)^{1/2}$$

The fourth moment is influenced by the small portion of inclusions that lie above the upper limit of a normal distribution (mean plus $3\sigma$). The fourth root of the fourth moment about zero is calculated by:

$$X_4 = (V_4)^{1/4}$$

For a normal distribution the ratio $X_4/\tilde{X}$ is 1.32. A higher value indicates the presence of some fields with an abnormally high inclusion content.

The moments about the mean values are utilized to calculate the standard deviation and two indicators of departure from a normal distribution —— skew and kurtosis. If a normal distribution is indicated, the confidence limit can be calculated. The moments about zero must be converted to moments about the mean. The calculation of the variance, $U_2$, is:

$$U_2 = V_2 - V_1^2$$

The standard deviation of the observations about the mean, $\sigma$, is calculated as:

$$\sigma = (U_2)^{1/2}$$

The standard deviation of the mean is then calculated as:

$$\sigma_{\overline{X}} = \frac{\sigma}{\sqrt{n-1}}$$

where n is the number of sets.

The third and fourth moments about the mean are calculated prior to determining skew and kurtosis:

$$U_3 = V_3 - 3V_1V_2 + 2(V_1)^3$$

$$U_4 = V_4 - 4V_1V_3 + 6(V_1)^2 V_2 - 3(V_1)^4$$

The skewness, $\beta_1$, is then calculated as:

$$\beta_1 = \frac{U_3 |U_3|}{(U_2)^3}$$

The closer $\beta_1$ is to zero, the better the data conform to a normal distribution. Kurtosis, $\beta_2$, is calculated as:

$$\beta_2 = \frac{U_4}{(U_2)^2}$$

For a large number of observations, $\beta_2$ approaches 3 for a Gaussian distribution. Values greater than 5 indicate non-normal distributions.

Examples of Inclusion Measurement and Data Analysis. We have followed Moore's procedure in our studies of inclusion assessment using the Leitz Texture Analyzer System with 16, 32 and 80X objectives. Nine samples containing sulfur contents from 0.020 to 0.34% sulfur were used. The sample types included air-melted, vacuum-degassed, as-cast, and wrought steels. In our study each set consisted of three rows of 30 contiguous fields, and 12 sets, i.e., 1080 fields, were measured. For these test conditions the total measured areas corresponded to 165.8, 42.69 and 6.63 sq mm for the 16, 32 and 80X objectives. All of the inclusions were evaluated for their volume fraction. The data are listed in Table 23 and plotted in Fig. 48. Except for two of the 80X measurements of inclusion content, all of the set data exhibited a normal distribution. The data were evaluated by means of linear regression analysis. The trend line shown is for all of the test data. The data were also examined with respect to the difference in range of set means, i.e., highest minus lowest set mean. Figure 49 plots the difference in the extremes of the set means against the sulfur content. The regression curves increase with increasing sulfur, and the increase is greater as the magnification rises. We normalized the data by dividing the difference in the extremes of the set means by the set mean and we expressed this number as a percent; we then plotted these values against the sulfur content (Fig. 50). The normalized data are seen to decrease as the sulfur content increases. As the magnification rises, the normalized value drops faster with rising inclusion content. The data presented in Figs. 49 and 50 show the importance of the increased area examined using the low-power objective. The test data can be summarized as follows:

| Objective | Range of % Relative Error (±) | | Average |
|-----------|-------------------------------|------|---------|
| 16X | 3.2 to | 7.5 | 5.1 |
| 32X | 3.9 to | 11.6 | 6.5 |
| 80X | 7.5 to | 23 | 11.9 |

## TABLE 23

### Results of image analysis measurement —— Bethlehem samples

| Sample | Sulfur (%) | Objective (X) | $A_A$ (%) | Error Bar, % (95% C.L.) | Rel. Error (%) | $\tilde{A}_A$ (%) | $A_{A_4}$ (%) | Skew (B1) | Kurtosis (B2) |
|---|---|---|---|---|---|---|---|---|---|
| EX33 | 0.020 | 16 | .265 | .245 – .285 | 7.5 | .267 | .272 | 1.14 | 3.35 |
| | | 32 | .277 | .245 – .309 | 11.6 | .282 | .293 | 1.156 | 3.4 |
| | | 80 | .23 | .2 – .26 | 13 | .236 | .246 | .01 | 2.56 |
| SO22-4 | 0.048 | 16 | .516 | .485 – .548 | 6.1 | .519 | .524 | -.52 | 3.7 |
| | | 32 | .478 | .446 – .509 | 6.6 | .48 | .486 | -.336 | 4.52 |
| | | 80 | .464 | .409 – .518 | 11.7 | .472 | .487 | -.067 | 2.64 |
| CO64 | 0.078 | 16 | .54 | .512 – .569 | 5.3 | .542 | .547 | 1.28 | 3.91 |
| | | 32 | .546 | .494 – .598 | 9.5 | .553 | .568 | 1.96 | 4.4 |
| | | 80 | .519 | .399 – .638 | 23 | .555 | .664 | 7.51 | 9.09 |
| SO25-4 | 0.090 | 16 | .906 | .858 – .954 | 5.3 | .91 | .917 | .063 | 2.025 |
| | | 32 | .763 | .733 – .793 | 3.9 | .764 | .767 | .0005 | 1.16 |
| | | 80 | .755 | .696 – .815 | 7.9 | .762 | .775 | -.172 | 3.3 |
| PO66-3 | 0.097 | 16 | .828 | .801 – .854 | 3.2 | .829 | .831 | .2 | 1.02 |
| | | 32 | .783 | .745 – .821 | 4.9 | .786 | .79 | .006 | 2.0 |
| | | 80 | .761 | .689 – .833 | 9.5 | .77 | .79 | .308 | 2.56 |
| CO65 | 0.13 | 16 | .765 | .732 – .799 | 4.4 | .767 | .771 | .085 | 2.8 |
| | | 32 | .729 | .676 – .782 | 7.3 | .734 | .746 | 1.4 | 4.14 |
| | | 80 | .700 | .619 – .782 | 11.6 | .714 | .739 | .15 | 2.45 |
| 1140 Mod. | 0.20 | 16 | 1.261 | 1.168 – 1.354 | 7.4 | 1.27 | 1.29 | -.03 | 1.45 |
| | | 32 | 1.243 | 1.174 – 1.313 | 5.6 | 1.25 | 1.26 | -.007 | 2.12 |
| | | 80 | 1.11 | .942 – 1.281 | 15.3 | 1.146 | 1.2 | -.376 | 2.445 |
| A1 | 0.30 | 16 | 1.878 | 1.812 – 1.944 | 3.5 | 1.88 | 1.89 | 0.06 | 2.99 |
| | | 32 | 1.99 | 1.898 – 2.088 | 4.8 | 2.0 | 2.01 | .55 | 2.93 |
| | | 80 | 1.855 | 1.716 – 1.995 | 7.5 | 1.87 | 1.9 | -.17 | 1.92 |
| B4 | 0.34 | 16 | 1.82 | 1.761 – 1.879 | 3.2 | 1.82 | 1.83 | -.01 | 2.85 |
| | | 32 | 2.00 | 1.905 – 2.095 | 4.8 | 2.01 | 2.02 | .16 | 1.12 |
| | | 80 | 2.00 | 1.847 – 2.16 | 7.8 | 2.02 | 2.05 | -2.88 | 24.27 |

Note:   All samples analyzed with 12 sets, 3 parallel rows of 30 contiguous fields, 1080 fields.

The relative accuracies using the 16 and 32X objectives are quite similar. An inspection of Fig. 48 shows that the 32X data generally lie closer to the trend line than do the 16X data (in 5 of 9 cases, the data are closer; in two cases about equal and in two cases the 16X data are closer). Hence, we have a slight preference for the 32X as opposed to the 16X objective.

The inclusion content of these samples was also measured by a standard point counting procedure and by lineal analysis using a Hurlbut counter. For the manual point count a magnification of 500X was used and a 100-point density grid was used to analyze 100 fields ($P_T$=10,000). The data were grouped into 10 sets, each containing 10 field measurements. This analysis required about one hour per sample. For the lineal analysis eight to ten measurements of the inclusion content were made using traverses (each about 15 minutes long) at 1000X in order to cover at least 10,000 units on the Hurlbut counter. These measurements were used in place of set means to calculate the accuracy. The resulting analysis data are listed in Table 24. All but one of the test data points conform to a normal distribution. Figure 51 is a plot of the manual point count data, and Fig. 52 is a plot of the lineal

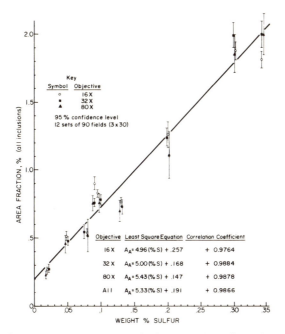

Fig. 48. Volume fraction measurements of nine samples of varying sulfur content using 16, 32 and 80X objectives. Line shown is least square fit for all of the data points.

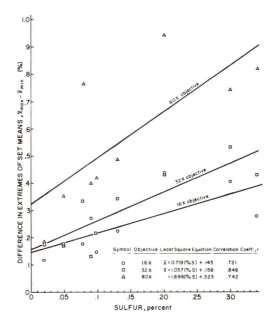

Fig. 49. Comparison of the extremes of set means for each measurement point shown in Fig. 48.

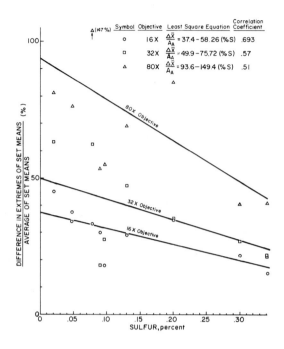

Fig. 50.   Data given in Fig. 49 normalized by set mean values in Fig. 48.

## TABLE 24
### Point count and lineal analysis data

Manual Point Count (10,000 points, 500X)

| Sample | %S | $A_A$ (%) | Error Bar, % (95% C.L.) | Rel. Error (%) | $\widehat{A_A}$ (%) | $A_{A_4}$ (%) | $B_1$ | $B_2$ |
|--------|------|-------|-----------------|------|-------|-------|--------|-------|
| EX33 | .020 | .22 | .127 - .313 | 42.4 | .261 | .307 | .0005 | .3 |
| SO22-4 | .048 | .43 | .23 - .63 | 46.5 | .524 | .657 | .534 | 2.273 |
| CO64 | .078 | .60 | .46 - .74 | 22.6 | .634 | .681 | -.12 | 1.772 |
| SO25-4 | .090 | 1.01 | .78 - 1.24 | 23.0 | 1.068 | 1.184 | .665 | 3.255 |
| PO66-3 | .097 | .80 | .61 - .99 | 24.2 | .851 | .95 | .661 | 2.55 |
| CO65 | .13 | .72 | .54 - .90 | 25.5 | .771 | .851 | .009 | 2.11 |
| 1140 Mod. | .20 | 1.10 | .69 - 1.51 | 37.1 | 1.26 | 1.487 | .211 | 1.77 |
| Al | .30 | 1.845 | 1.74 - 1.95 | 5.9 | 1.909 | 2.02 | .002 | 2.14 |
| B4 | .34 | 1.74 | 1.59 - 1.89 | 8.6 | 1.867 | 2.04 | -.095 | 2.02 |

Lineal Analysis (1000X)

| Sample | %S | $A_A$ (%) | Error Bar, % (95% C.L.) | Rel. Error (%) | $\widehat{A_A}$ (%) | $A_{A_4}$ (%) | $B_1$ | $B_2$ |
|--------|------|-------|-----------------|------|-------|-------|--------|-------|
| EX33 | .020 | .23 | .16 - .3 | 31.2 | .24 | .27 | .16 | 2.23 |
| SO22-4 | .048 | .434 | .3 - .57 | 31.1 | .47 | .53 | .4 | 1.91 |
| CO64 | .078 | .32 | .24 - .41 | 26.1 | .34 | .37 | -.108 | 13.96 |
| SO25-4 | .090 | .72 | .44 - 1.0 | 38.5 | .79 | .88 | .002 | 1.5 |
| PO66-3 | .097 | .8 | .58 - 1.02 | 27.6 | .84 | .90 | .00002 | 1.93 |
| CO65 | .13 | .88 | .58 - 1.18 | 34 | .98 | 1.12 | .11 | 1.95 |
| 1140 Mod. | .20 | 1.37 | .84 - 1.89 | 38.2 | 1.53 | 1.78 | .17 | 2.02 |
| Al | .30 | 1.69 | 1.38 - 2.0 | 18.5 | 1.74 | 1.83 | -.25 | 1.5 |
| B4 | .34 | 1.76 | 1.51 - 2.0 | 13.9 | 1.79 | 1.87 | .45 | 1.78 |

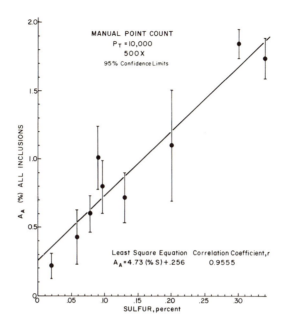

Fig. 51. Inclusion volume fractions of the nine samples of varying sulfur content evaluated by the manual point count method.

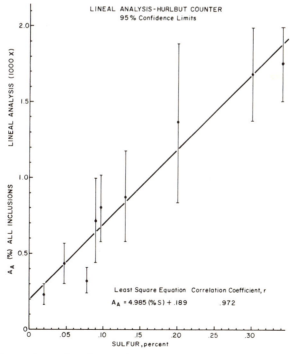

Fig. 52. Inclusion volume fractions of the nine samples of varying sulfur content evaluated by the lineal analysis method using a Hurlbut counter.

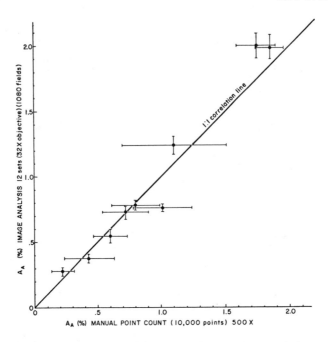

Fig. 53. Comparison of the mean and 95% confidence limits of the volume fraction measurements obtained by image analysis (32X objective) and by point counting.

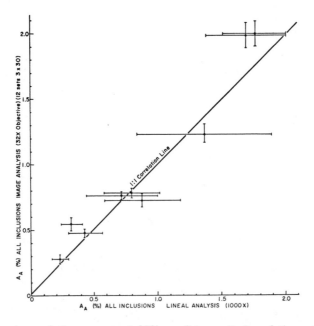

Fig. 54. Comparison of the mean and 95% confidence limits of the volume fraction measurements obtained by image analysis (32X objective) and by manual lineal analysis.

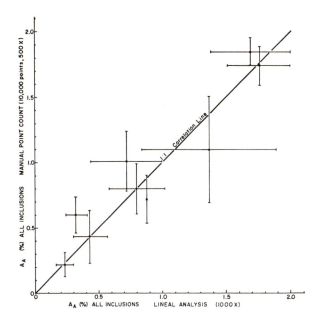

Fig. 55. Comparison of the mean and 95% confidence limits of the volume fraction measurements obtained by manual point count and by manual lineal analysis.

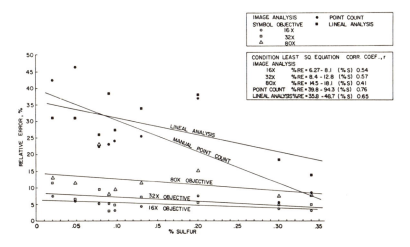

Fig. 56. Plot of the percent relative error for the nine samples of varying sulfur content using image analysis (16, 32 and 80X objectives), manual point counting and manual lineal analysis.

## TABLE 25
## Summary of inclusion detection and measurement methods — possible information obtainable by methods

| Methods | Inclusion Size Range Normally Detected | | Manner of Quantification of Inclusion Content | | | Information on Size and Distribution | Information Regarding Shape | Nature of Inclusion Identification | | |
|---|---|---|---|---|---|---|---|---|---|---|
| | Microscopic | Macroscopic | Volume % | Weight % | Semi-Quantitative Index | | | Optical | Chemical Elements | Formula of Compound |
| **A) Optical Microscopy** | | | | | | | | | | |
| o Comparison Charts | X | (1) | | | X | No | Yes | X | | |
| o Counting Methods | X | (1) | X | | | Possible with some methods(2) | Yes | X | | |
| **Stereologically-based Methods** | | | | | | | | | | |
| o Point Counting | X | (1) | X | | | No | No | X | | |
| o Lineal Analysis | X | (1) | X | | | Size - chord length Distribution - mean free path | Yes | X | | |
| o Image Analysis | X | (1) | X | | | Yes | Yes | X | | |
| **B) Macroscopic** | | | | | | | | | | |
| o Magnetic Particle | | X | | | X | Yes | Yes(3) | No information provided | | |
| o Fracture | | X | | | X | Yes | Yes(3) | No information provided | | |
| o Hot Acid Etch | X(5) | X | | | X | No | No(1) | No information provided(4) | | |
| o Contact Printing | | X | | | X | No | No(1) | Qualitative for some elements | | |
| o Ultrasonics | X(2) | X | | | | Possible(6) | | No information provided | | |
| **C) Chemical** | | | | | | | | | | |
| o Oxygen Analysis | X | (1) | | X | | No | No | | X | |
| o Sulfur Analysis | X | (1) | | X | | No | No | | X | |
| o Extraction - residue analysis | X | (1) | | X | | Possible(7) | No | x(8) | X | X |
| **D) Other Methods** | | | | | | | | | | |
| o Microradiography | X | | x(9) | | | Possible(9) | Possible(9) | | | |
| o Physical Extraction - petrographic/x-ray analysis | X | X | | | | No(10) | No(10) | X | X | |
| o Electron Microprobe | X(11) | X(11) | x(9) | | | Possible(12) | Possible(12) | | X | x(13) |
| o Scanning Electron Microscopy | X(11) | X(11) | x(9) | | | Possible(12) | Possible(12) | | X | x(13) |
| o Replication of Failures - electron microscope examination | X | X | X | | | Yes | Yes | No information provided(14) | | |

1) Very low probability of detection but possible.
2) Capable but extent limited.
3) Detects elongated inclusions primarily, length and thickness can be measured.
4) 1 to 1 HCl/water etch attacks sulfides.
5) Microscopic inclusions contribute to image but individual features not resolvable.
6) Size information possible using immersion tank test methods.
7) Possible with Coulter counter measurements.

8) Residue can be examined in petrographic microscope.
9) Stereological methods can be applied.
10) Only individual inclusions are extracted but they can be measured with petrographic microscope.
11) Generally used to examine inclusions of specific interest which may be of microscopic or macroscopic size (assumes SEM has analytical capability).
12) It is possible to generate such data but not generally done.
13) Compound formula can be calculated from elemental analysis.
14) Analysis of extracted inclusions may be possible by electron diffraction.

analysis data. The width of the error bars in each case is much greater than that of the image analysis data. The average relative accuracies for the point count data and the lineal analysis data were 26.1% and 28.8%, respectively. In spite of the much longer time for lineal analysis as opposed to point count, the accuracy of the lineal analysis was not as good.

The image analysis data are compared with the point count data and the lineal analysis data as given in Figs. 53 and 54, respectively. The point count data are compared with the lineal analysis data in Fig. 55. As a final comparison, Fig. 56 gives a plot of the relative accuracy versus sulfur content for all of the data. This plot shows the strong dependency of the relative accuracy on the inclusion volume fraction for the manual techniques and a very minor influence for the image analysis data.

## SUMMARY

Because of the often harmful influence of inclusions on steel properties and behavior considerable effort has been expended by many researchers on efforts to measure the inclusion content of steel. Many different techniques, both macroscopic and microscopic, have been developed over the years. Table 25 summarizes data on these techniques. In general, all methods have certain key limitations, and it is recommended that in any study of inclusion content both macroscopic and microscopic tests be used. The macroscopic procedures, especially ultrasonic inspection, are useful to detect the worst conditions. These should then be studied by microscopic methods to determine the extent and nature of the macroscopic indications. As the severity of the intended application of the steel increases, the degree and amount of testing must also be increased.

The use of such tests, if they are to be of real value, must be firmly based on sound statistical procedures for sampling and data analysis. Sample preparation must level up to a number of requirements if results are to be meaningful. The present review has focused on many of the practical problems involved in inclusion measurements and evaluation. It is hoped that the information assembled and the extensive bibliography will be helpful not only for the actual measurement techniques, but for interpretation and statistical treatment of the resulting data.

## ACKNOWLEDGEMENTS

The author wishes to acknowledge the help of A.V. Brandemarte and S.A. Yagielski for image analysis measurements, W.A. Beveridge for art work, and B.S. Mikofsky for editorial review.

## REFERENCES

1. S. Bergh, "Oxygen Content and Oxides in a Silicon Killed Steel," *Jern. Ann.,* V. 146, 1962, p. 748–762 (BISI No. 3140).
2. G.W. Walker, "Rating of Inclusions ("Dirt Chart")," *Metal Progress,* V. 35, (Feb. 1939), p. 169, 170, 167.
3. H. Diergarten, "Slag Inclusions in Steel. Standards for Size and Frequency," *Metal Progress,* V. 32, Sept. 1937, p. 269–271.
4. H. Diergarten, "Classification of Roller Bearing Steels According to Inclusion Content," *Arch. Eisenhutten.,* V. 10, 1936, p. 197–210 (HB No. 556).
5. ASTM A317, "Standard Method of Macroetch Testing and Inspection of Steel Forgings"

6.  MIL—STD—1459 AC MU, "Macrograph Standards for Steel Bars, Billets, and Blooms for Ammunition Components," Sept. 4, 1973 (Superseding MIL—STD—1459 (MU) which superseded MIL—STD—430A)
7.  ASTM E381, "Standard Method for Rating Macroetched Steel"
8.  ASTM A561, "Standard Recommended Practice for Macroetch Testing of Tool Steel Bars"
9.  ASTM A604, "Standard Method for Macroetch Testing of Consumable Electrode Remelted Steel Bars and Billets"
10. ASTM E45, "Standard Recommended Practice for Determining the Inclusion Content of Steel"
11. ISO 3763, "Wrought Steels —— Macroscopic Methods for Assessing the Content of Non-Metallic Inclusions"
12. SIS 111110E, "Determination of Slag Inclusion Content of Steel. Macroscopic Methods"
13. J.B. Morgan, "Ultrasonics Can Rate Cleanliness, Fatigue Life of Steel," *Metals Eng. Quart.,* V. 13, (Feb. 1973), p. 1—4.
14. J. Dumont-Fillon and M. Lacroix, "Application of Ultrasonics to the Comparison of the Inclusion Contents of Steels," *Mem. Sci. Rev. Met.,* V. 75, No. 8/9, (Aug.—Sept. 1978), p. 491—501.
15. ASTM E588, "Standard Recommended Practice for Detection of Large Inclusions in Bearing Steel by the Ultrasonic Method"
16. W.H. Burr, "Ultrasonic Rating of Bearing Steel Cleanliness," *ASTM STP 575,* )1975), p. 178—188.
17. S.S. Daniel and R.A. Rege, "Ultrasonic Cleanliness Rating of Steel," *J. Metals,* V. 23, (July 1971), p. 26—37.
18. W.W. Bayre and D.D. McCormack, "Ultrasonic Detection of Inclusions in Steel," *Materials Evaluation,* V. 28, (Feb. 1970), p. 25—31.
19. W.W. Bayre and D.D. McCormack, "Ultrasonic Detection of Inclusions in Bars, Tubes, and Billets," *Mech. Working and Steel Proc.,* VI, AIME, (1969), p. 111—134.
20. R.N. Cressman and A.J. Plante, "Ultrasonic Inclusion Rating of Bearing Steel Billets," *Blast Furn. and Steel Plant,* V. 57, No. 3, (March 1969), p. 232—235, 238, 251, 259.
21. C. Benedicts and B. Lofquist, *Non-Metallic Inclusions in Iron and Steel,* J. Wiley & Sons, NY, (1931).
22. R. Kiessling and N. Lange, *Non-Metallic Inclusions in Steel, Part I,* ISI SR90, London, 1964; Part II, ISI SR100, London, (1966).
    R. Kiessling, *Non-Metallic Inclusions in Steel, Part III,* ISI SR 115, London, 1968; R. Kiessling and N. Lange, *Non-Metallic Inclusions In Steel,* Parts I—IV, The Metals Society, 2nd ed., (1978).
23. T.R. Allmand, *Microscopic Identification of Inclusions in Steel,* Brit. Iron and Steel Res. Assoc., London, (1962).
24. L.E. Grant, "Identification of Inclusions in Steels," *Trans. ASST,* V. 19, (Dec. 1931), p. 165—181.
25. R. Graham and R. Hay, "Identification of Non-Metallic Inclusions in Steel," *J. Royal Tech. College, Glasgow,* V. 4, (1937), p. 77—84.
26. W. Campbell and G.F. Comstock, "Identification of Non-Metallic Inclusions in Iron and Steel," *Proc. ASTM,* V. 23, (1923), p. 521—522.
27. J.H. Whitely, "A Method of Identifying Manganese-Sulphide Inclusions in Steel," *JISI,* V. 160, (Dec. 1948), p. 365—366.
28. R.G. Wells and J.M. Snook, "Petrographic Tools and Techniques in the Identification of Nonmetallic Inclusions in Steels," *Proc. Electric Furn. Conf.,* V. 18, (1960), p. 60—70.
29. T. Ernst and G. Konig, "A New Method for the Determination of Nonmetallic Inclusions in Steel," *Mitteilungen,* V. 7, No. 12, (1959), p. 93—96 (BISI No. 1752).

30. W.D. Forgeng, "Techniques for the Study of Nonmetallic Inclusions," *Proc. Electric Furn. Conf.,* V. 18, (1960), p. 7–21.

31. F.F. Franklin, "A Study of Inclusions in Steel. Part II, Metallographic Examination," *Proc. Electric Furn. Conf.,* V. 20, (1962), p. 100–124.

32. R. Kusaka et al, "Nonmetallic Inclusions in Steel Studied by the Use of Microscopic Thin Sections," *Tetsu-to-Hagane,* V. 50, No. 3, (1964), p. 494–496 (HB No. 6385).

33. M. Sasaki et al, "Identification of Nonmetallic Inclusions by Polished Thin Sections," *Tetsu-to-Hagane,* V. 5, No. 4, (Dec. 1965), p. 324–329.

34. H.J. Schrader, "Microscopic Identification of Nonmetallic and Metallic Inclusions in Steels," *Prakt, Met.,* V. 4, No. 9, (Sept. 1967), p. 470–480.

35. K.H. Mertins, "The Metallographic Identification of Non-Metallic Inclusions in Steels," *Prakt. Met.,* V. 5, No. 12, (1968), p. 669–685.

36. R.B. Snow, "Identification of Oxides, Silicates, and Aluminates in Steel by Optical Methods," *J. Metals,* V. 20, (Oct. 1973), p. 75–77.

37. W.G. Wilson and R.G. Wells, "Identification of Inclusions in Rare Earth Treated Steels," *Metal Prog.,* V. 104, (Dec. 1973), p. 75–77.

38. T.R. Allmand and D.H. Houseman, "Thin Film Interference –– A New Method for Identification of Non-Metallic Inclusions," *Microscope,* V. 18, (1970), p. 11–23.

39. F.A. Silber and H. Judtmann, "Vapour Phase Deposited Interference Layers as a Means of Identifying Inclusions," *Prakt. Met.,* V. 9, No. 6, (June 1972), p. 318–328.

40. R.G. Wells, "Metallographic Techniques in the Identification of Sulfide Inclusions in Steel," *Sulfide Inclusions in Steel,* ASM, (1975), p. 123–134.

41. A.M. Hall and E.E. Fletcher, "The Application of Color Photography to the Study of Nonmetallic Inclusions," *ASTM STP 86,* (1949), p. 59–63.

42. M. Sarracino and F.S. Rossi, "Identification and Location of Small Lead Particles Dispersed in As-Cast Stainless Steel," *Prakt. Met.,* V. 8, No. 5, (May 1971), p. 300–308.

43. G.M. Chalfant, "Revealing Lead Inclusions in Leaded Steels," *Metal Progress,* V. 78, (Sept. 1960), p. 77–79.

44. H. Morrogh, "The Examination and Identification of Inclusions in Metals and Alloys," *Polarized Light in Metallography,* Butterworths, London, (1952), p. 88–104.

45. M.A. Scheil et al, "Identification of Inclusions in Steel by the Use of Reflected Polarized Light," *Trans. ASM,* V. 27, (June 1939), p. 479–504.

46. S.L. Hoyt and M.A. Scheil, "Use of Reflected Polarized Light in the Study of Inclusions in Metals," *Trans. AIME,* V. 116, (1935), p. 405–424.

47. F. Hartman, "Investigations on Inclusions in Steel with a Polarizing Microscope," *Arch. Eisenhutt.,* V. 4, (1931), p. 601–606.

48. "Identification of Inclusions in Steel by the Use of Reflected Polarized Light," *ASM Metals Handbook,* 6th ed., (1939), p. 741–749.

49. G.L. Kehl et al, "The Removal of Inclusions for Analysis by an Ultrasonic 'Jack Hammer'," *Metallurgia,* V. 55, (March 1957), p. 151–154.

50. W.J. McGonnagle, "Ultrasonic Jack Hammer Developments," *Metallurgia,* V. 65, (April 1962), p. 205–206.

51. I. Uchiyama et al, "Extraction of Nonmetallic Inclusions from Steel by Using an Ultrasonic Drill, " *Tetsu-to-Hagane,* V. 47, No. 3, (1961), p. 519–521 (HB No. 6063).

52. W. Betteridge and R.S. Sharpe, "The Study of Segregations and Inclusions in Steel by Micro-Radiography," *JISI,* V. 158, (Feb. 1948), p. 185–191.

53. J.R. Blank and W. Johnson, "Nature and Distribution of Inclusions in Leaded Steels," *Steel Times,* V. 191, (July 21, 1965), p. 110–115; (July 30, 1965), p. 148–152; (August 6, 1965), p. 176–181.

54. A. Podgornik and A. Smolej, "The Determination of Size, Shape and Distribution of Chip Breaking Inclusions in Light Metal Free-Cutting Alloys by Microradiography," *Prakt. Met.*, V. 8, No. 2, (Feb. 1971), p. 113—117.

55. *Determination of Nonmetallic Compounds in Steel, ASTM STP 393*, (1966).

56. T.E. Rooney, "The Alcoholic Iodine Method for the Separation of Oxides in Steel," *ISI SR 25, Pt 6A,* (1939), p. 141—158.

57. S. Bergh and A. Josefsson, "A Simplified Method for Electrolytic Isolation of Slag Inclusions in Steels," *Jern. Ann.,* V. 145, (1961), p. 551—560 (HB No. 5424).

58. R.G. Smerko and D.A. Flinchbaugh, "Recent Progress in the Chemical Extraction of Nonmetallic Inclusions in Steel — Techniques and Application," *J. Metals,* V. 20, (July 1968), p. 43—51.

59. W.M. Wojcik et al, "Inclusion Counting of Steel Residues as a Means of Assessing the Hot Working Performance of Steel," *J. Metals,* V 19, (Dec. 1967), p. 36—40.

60. H. Walz and R.A. Bloom, "Non-Metallic Inclusions — Their Quantitative Extraction From Steel," *J. Metals,* V. 12, (Dec. 1960), p. 928—932.

61. N.I. Iwamoto, "Oxide Inclusions Formed in Steels (Report I) — Deoxidation Products by Al, Si, Mn, Ti, V," *Trans. JWRI,* V. 3, No. 1, (1974), p. 41—51.

62. K. Narita, "Observation, Identification and Determination of Nonmetallic Inclusions and Precipitate in Steel," *Trans. ISIJ,* V. 16, (1976), p. 208—213.

63. K.E. Burke, "Chemical Extraction of Refractory Inclusions from Iron- and Nickel-Base Alloys," *Metallography,* V. 8, (1975), p. 473—488.

64. M. Morait and E. Mustacescu, "The Tentative Establishment of Some Correlations Between the Electrolytic Dissolution Method and the Metallographic Method of Determining the Non-Metallic Inclusions in Steel," *Metallurgia,* (1966), V. 18, No. 3, p. 154—156.

65. K. Segawa, "Recent Developments in the Isolation of Inclusions in Steel," *Tetsu-to-Hagane,* V. 52, (June 1966), p. 967—980.

66. A. More, "Isolation and Ease of Isolating Oxide Inclusions in Austenitic Chromium-Nickel Steels," *Arch. Eisenhutten.,* V. 37, (June 1966), p. 473—481.

67. H. Kroll and P. Lenk, "Isolation of Oxide Inclusions from Steel," *Neue Hutte,* V. 12, (July 1967), p. 434—438.

68. M. Ihida et al, "Study of Some Problems on the Determination of Oxygen and the Isolation of Oxide Inclusions in Iron and Steel," *Nippon Kokan Tech. Rep.,* V. 41, (1967), p. 405—415.

69. P. Dickens and P. Konig, "Isolation of Inclusions in Steel by Direct Chlorination," *Arch. Eisenhutt.,* V. 39, (June 1968), p. 453—456.

70. K. Narita et al, "A Fundamental Study of the Isolation and Determination of Oxide Inclusions in Steel by Nitric Acid, Iodine-Alcohol, Bromine-Ester and Chlorination Methods," *Tetsu-to-Hagane,"* V. 55, No. 9, (August 1969), p. 846—860.

71. A.S. Korchemkina et al, "Factors Affecting the Completeness of the Separation of Sulphide Inclusions from Steel by the Electrolysis Method," *Zavods. Lab.,* (1966), p. 1324—1327.

72. S. Maekawa and Y. Shiga, "Determination of Sulphide in Steel by the Electrolytic Method," *Tetsu-to-Hagane,* V. 55, No. 13, (Nov. 1969), p. 1263—1269.

73. G. Schmolke, "Separation of Sulphides from Plain Carbon Steels," *Arch. Eisenhutten.,* V. 46, No. 4, (April 1975), p. 261—264.

74. D.A. Flinchbauch, "Use of a Modified Coulter Counter for Determining Size Distribution of Macroinclusions Extracted from Plain Carbon Steels, " *Anal. Chem.,* V. 43, (Feb. 1971), p. 178—182.

75. D.A. Flinchbauch, "Determining Size Distribution of Oxides in Plain Carbon Steels by Halogen-in-Organic Solvent Extraction and Coulter Counter Measurement," *Anal. Chem.,* V. 41, (Dec. 1969), p. 2017—2023.

76. O. Kammori et al, "An Ultrasonic Sieving Method for Classification of Precipitates and Inclusions Extracted from Steels According to Size," *Nippon Kinzoku,* V. 33, No. 6, (1969), p. 669–672.

77. E. E. Wicker, "Activation Analysis," *Determination of Gaseous Elements in Metals,* J. Wiley & Sons, (1974), p. 75–111.

78. W.R. Bandi et al, "Iron and Steel," *Determination of Gaseous Elements in Metals,* J. Wiley & Sons, (1974), p. 501–551.

79. J.F. Martin and L.M. Melnick, "Vacuum and Inert-Gas Fusion," *Determination of Gaseous Elements in Metals,* J. Wiley & Sons, (1974), p. 113–219.

80. T.J. Baker and J.A. Charles, "Deformation of MnS Inclusions in Steel," *JISI,* V. 210, (Sept. 1972), p. 680–690.

81. W. Dahl et al, "Behavior of Different Types of Sulphide During Deformation and Their Effect on the Mechanical Properties," *Stahl. Eisen,* V. 86, (June 30, 1966), p. 796–817 (BISI No. 5010).

82. S. Ekerot and B.I. Klevebring, "A Note on the Behavior of Silicate Inclusions During Hot Working," *Scand. J. Met.,* V. 3, (1974), p. 151–152.

83. P.E. Waudby, "Factors Controlling the Plasticity of Silicate Inclusions," *Steel Times Annual Review,* (1972), p. 147–152.

84. T. Fastner and W. Schwenzfeier, "Behavior of Macroscopic Inclusions During Rolling," *BHM,* V. 9, (1974), p. 353–361 (BISI No. 12841).

85. P.J.H. Maunder and J.A. Charles, "Behavior of Non-Metallic Inclusions in a 0.2% Carbon Steel Ingot During Hot Rolling," *JISI,* V. 206, (July 1968), p. 705–715.

86. S. Ekerot, "Behavior of Slag Inclusions of Different Composition During Hot Working Conditions," *Clean Steel,* V. 1, Report of the Royal Swedish Academy of Eng. and Sci., 169;1, Stockholm, (1971), p. 217–227.

87. B.I. Klevebring, "The Deformation of Non-Metallic Inclusions in Steel During Hot Working," *Scand. J. Met.,* V. 3, (1974), p. 102–104.

88. K.B. Gove and J.A. Charles, "Further Aspects of Inclusion Deformation," *Metals Tech.,* (Sept. 1974), p. 425–431.

89. T. Malkiewicz and S. Rudnik, "Deformation of Non-Metallic Inclusions During Rolling of Steel," *JISI,* V. 201, (Jan. 1963), p. 33–38.

90. J.C. Brunet and J. Bellot, "Deformation of MnS Inclusions in Steel," *JISI,* V. 211, (July 1973), p. 511–512.

91. T.R. Allmand, "A Review of Methods for Assessing Nonmetallic Inclusions in Steel," *JISI,* V. 190, (Dec. 1958), p. 359–372.

92. T.R. Allmand and D.S. Coleman, "The Effect of Sectioning Errors on Microscopic Determinations of Non-metallic Inclusions in Steels," *Metals and Materials,* V. 7, (June 1973), p. 280–283.

93. B. Rinman et al, "Inclusion Chart for the Estimation of Slag Inclusions in Steel," Jernkontoret, Stockholm, Sweden, Uppsala, (1936), 24 pages.

94. B. Rinman et al, "Chart for the Estimation of Inclusions in Steel," *Jern. Ann.,* V. 120, (1936), p. 199–226.

95. G.R. Bolsover, "Non-Metallic Inclusions in Steel," *Metallurgia,* (July 1935), p. 83–84.

96. "The 'Fox' Inclusion Count. A Quantitative Method of Expressing the Cleanness of Steel," S. Fox & Co. Ltd.

97. SAE Recommended Practice, "Microscopic Determination of Inclusions in Steels, –– SAE J422a," SAE Handbook, Supplement HS30, (1971), p. 98–100.

98. German Standard VDEh 1570–71, "Microscopical Examination of Special Steels for Non-Metallic Inclusions Using Standard Micrograph Charts," *Stahl–Eisen Prufblatt 1570,* 2nd ed., (Aug. 1971)(BISI 10905).

99. ISO Technical Committee 17–Steel Sub-Committee 7, "Proposal of Germany for an ISO Standard Concerning the Microscopic Examination of Specialty Steels for Non-Metallic Inclusions with Inclusion Charts," (May 1972).

100. German Standard 1572, "Microscopic Testing of Free-Cutting Steels for Non-Metallic Sulphide Inclusions by Means of a Series of Pictures," *Stahl-Eisen Prufblatt 1572*, Issue 1, (Sept. 1977) (BISI 15913).

101. Italian Standard UNI 3244—66, "Microscopic Examination of Ferrous Materials. Methods for Assessment of Non-Metallic Inclusions in Steels," (Oct. 1966) (BISI 12242).

102. A. Koyanagi and M. Kinoshi, "Several Rating Methods of Nonmetallic Inclusions in Bearing Steels in Japan," *ASTM STP 575*, (1975), p. 22—37.

103. Japanese Standard JIS-G—0555, "Microscopic Testing Method for the Non-metallic Inclusions in Steel," *JIS Handbook*, (1965).

104. USSR Standard GOST 1778—70, "Metallographic Methods of Determination of Nonmetallic Inclusions," Moscow, (1970), (BISI 16703).

105. USSR Standard GOST 801—60, "Standard for Ball and Roller Bearing Steel," Moscow, (1960).

106. S. Alisanova, "Method of Numerical Evaluation of Inclusion Content of Steel," *Zav. Lab.*, V. 10, No. 5, (May 1941), p. 521—522, (HB No. 1499).

107. D.E. Nulk, "A User Looks at . . . Cleanliness of Vacuum-Melted Alloys," *Metal Progress*, V. 74, (Aug. 1958), p. 103—109.

108. D. Schreiber and H. Ziehm, "Quantitative Evaluation of a Classification Series for Non-Metallic Inclusions in Free Cutting Steels," *Prakt. Met.*, V. 6, No. 10, (Oct. 1969), p. 585—595.

109. D. Ruhl, "A Counting Out Technique for the Quantitative Determination of the Degree of Purity of Steels," *Prakt. Met.*, V. 6, No. 10, (Oct. 1969), p. 603—613.

110. E. Scheil, "Statistical Studies of Structure," Part 1, *Z. Metallk.*, V. 27, (1935), p. 199—209 (Summary in HB 268).

111. Y.N. Malinochka and B.P. Moiseev, "Metallographic Method of Evaluating Cleanliness of Rimmed Steel," *Ind. Lab.*, V. 42, No. 6, (June 1976), p. 909—913.

112. Georg Schafer & Co., "Recommended Practice for Determining the Inclusion Content of Roller Bearing Steels," Report W1—67, (March 1, 1967).

113. E. DiGianfrancesco and P. Filippi, "Use of Automatic Analysis of the Image for Determining Morphological Parameters of Manganese Sulphides in Steels with a High Standard of Machinability," *Met. Italiana*, (1976), No. 9, p. 434—438 (BISI No. 17370).

114. H.J. Eckstein and K. Schneider, "The Evaluation of Testing Methods for Non-Metallic Inclusions in Carbon Steel — CK 22," *Neue Hutte*, V. 5, (Oct. 1960), p. 607—617.

115. J.J.C. Hoo, "Reexamination of Rating Methods of Nonmetallic Inclusions in Bearing Steels," *ASTM STP 575*, (1975), p. 38—48.

116. K. Barteld and A. Stanz, "Microscopic Examination of Specialty Steels for Non-Metallic Inclusions with Reference Inclusion Charts," *Arch. Eisenhutt.*, V. 42, No. 8, (Aug. 1971), p. 581—597.

117. W.M. Wojcik and H.F. Walter, "A Method for Quantitative Evaluation of Steel Cleanliness," *J. Metals*, V. 18, (June 1966), p. 731—735.

118. J.V. Hardy and R.T. Allsop, "Counting Non-Metallic Inclusions in Steel," *JISI*, V. 195, (July 1960), p. 302—306.

119. J.R. Blank and T.R. Allmand, "Evaluation of Operator Errors Occurring in Assessment of Non-metallic Inclusions by Conventional Metallographic Methods," *Automatic Cleanliness Assessment of Steel*, ISI SP 112, (1968), p. 1—18.

120. H.J. Spies, "Influence of the Conditions of Testing on the Results of Microscopic Cleanness Determinations," *Neue Hutte*, V. 11, (July 1966), p. 420—427 (BISI 5195).

121. H.J. Spies, "Metallographic Methods for the Determination of the Degree of Purity in Steel," *Prakt. Met.*, V. 3, No. 4, (May 1966), p. 171—183.

122. M.S. Mikhalev and L.V. Mironov, "Metallographic Control of Steel According to Non-metallic Inclusions," *Stal in English,* V. 20, No. 7, (1960), p. 519–521.

123. W.H. Hatfield and G.W. Giles, "Non-Metallic Inclusions in Steel. Quantitative Evaluation, Part I," *JISI,* V. 42, (1940), p. 237P–276P.

124. "Report of the Inclusions Count Sub-Committee," *ISI SR 25,* Eight Report on the Heterogeniety of Steel Ingots, (1939), p. 305–322.

125. S. Epstein, "A Suggested Method of Determining the Cleanness of a Heat of Steel," *Metals and Alloys,* V. 2, No. 4, (Oct. 1931), p. 186–191.

126. A.B. Kinzel and W. Crafts, "Inclusions and Their Effect on Impact Strength of Steel," *Trans. AIME,* TP402, (Feb. 1931), p. 143–195.

127. H. Kjerrman, "Method for Determination of Nonmetallic Inclusions in Steel and Some Results Obtained with It," *Jern. Ann.,* V. 113, No. 4, (1929), p. 181–199 (HB No. 76).

128. C.H. Herty, Jr. et al, "The Physical Chemistry of Steel Making; Deoxidation with Silicon in the Basic Open Hearth Process," *Cooperative Bulletin No. 38,* Bureau of Mines and Bureau of Met. Res., Carnegie, Inst. of Tech., (1930), p. 25–29.

129. B. Wilandh, "Microscopic Slag Determination by the Picture-Drawing Method," *Jern. Ann.,* V. 149, No. 4, (1965), p. 156–164.

130. S. Bergh and O. Lindberg, "Routine Appraising of the Micro-slag Contents in Steel by the Size-Measurement —— Point Counting Method," *Jern. Ann.,* V. 149, No. 4, (1965), p. 150–156.

131. W. Zieler, "The Nonmetallic Inclusions in Steel, Their Distribution in the Cast Ingot and the Way in Which They are Affected by Deoxidation Agents, Especially by Zirconium, Sodium and Calcium," *Arch. Eisenhutt.,* V. 5, (Dec. 1931), p. 299–314.

132. R. Perrin, "Perrin's Rapid Steel Refining," *The Iron Age,* V. 140, (Oct. 14, 1937), p. 123–148.

133. R. Hunter, "Determination of Nonmetallic Inclusions in Steel," *Metal Treatment,* V. 4, (1938), p. 177–181.

134. J.H. Whiteley, "Inclusions in a Series of Bath Samples from an Electric Furnace and a Note on Sulphides," Seventh Report on the Heterogeneity of Steel Ingots, *ISI SR 16,* (1937), p. 23–55.

135. T. Swinden, "Rimming Steel. Further Studies on the Composition Variation from Outside to Centre," Ninth Report on the Heterogeneity of Steel Ingots, *ISI SR 27,* (1939), p. 17–41.

136. E. Maurer and R. Schustek, "Methods for Determining Cleanliness of Steel," *Stahl Eisen,* V. 63, No. 40, (1943), p. 725–727 (HB No. 1897).

137. A.D. Baynes, "An Inclusion Rating for Aluminum-Deoxidized Plain Carbon Cast Steels," *British Foundryman,* V. 56, No. 1, (Jan. 9, 1963), p. 9–18.

138. A.D. Baynes, "Influence of Deoxidation Practice in Basic Electric Arc Steelmaking on the Cleanness and Mechanical Properties of 0.20–0.25% Carbon Cast Steels," *British Foundryman,* (Nov. 1966), p. 451–462.

139. J. Welchner and W.G. Hildorf, "Relationship of Inclusion Content and Transverse Ductility of a Chromium-Nickel-Molybdenum Gun Steel," *Trans. ASM,* V. 42, (1950), p. 455–485.

140. M.S. Aronovich and I.M. Lyubarskii, "Determination of Nonmetallic Inclusions in Steel with a Microscope," *Metallurg,* V. 11, No. 9, (1936), p. 89–95.

141. J. Czochralski, "A Method for the Quantitative Definition of Nonmetallic Inclusions in Metals and Alloys," *Wiadomosci Inst. Metalurg. Metaloznawslwa,* V. 2, (1935), p. 34–37.

142. "Determination of Inclusions. SAE Recommended Practice," *SAE Handbook,* 1942 ed., p. 334–339, 1954 ed., p. 160–163.

143. "Classification of Inclusions in Steel," *SAE Journal,* V. 46, No. 6, (June 1940), p. 17–19.

144. BISRA Non-metallic Inclusion Group, "A Study of the Lineal Traverse Method for Counting Inclusions in Steel," *JISI*, V. 204, (Feb. 1966), p. 146–151.

145. T.R. Allmand, "Lineal Traverse Count for Assessing Abrasive Inclusion Content of High Sulphur Free-Machining Steels," *Steel Times*, V. 196, (Aug. 1968), p. 517–520.

146. S. Bergh, "Quantitative Microscopic Determination of the Inclusions in Steel," *Jern. Ann.*, V. 146, (1962), p. 924–934 (HB No. 5980).

147. O. Kammori et al, "Determination of Oxide Inclusions in Rimmed Steel Ingots After Classification According to Size," *J. Jpn. Inst. of Metals*, V. 8, (1968), p. 773–778 (BISI 8946).

148. Z. Buzek and V. Schindlerova, "Giving Precision to SIMS Classification of Inclusions Formed in Steels for Casting," *Slevarenstvi*, (1972), No. 10, p. 409–411 (BISI 11033).

149. S. Morykwas, "Inclusion Control Strengthens Investment Castings," *Metal Progress*, V. 93, (March 1968), p. 115, 116, 118.

150. D.J. Carney and E.C. Rudolphy, "Examination of a High Sulphur Free-Machining Ingot, Bloom and Billet Sections," *J. of Metals*, V. 5, (Aug. 1953), p. 999–1008.

151. L.H. Van Vlack, "Correlation of Machinability with Inclusion Characteristics in Resulphurized Bessemer Steels," *Trans. ASM*, V. 45, (1953), p. 741–757.

152. G.K. Bandyopadhyay et al, "Variations in the Average Size and Quantity of Nonmetallic Inclusions During Oxidation and Deoxidation of Steel," *Met. Trans.* V. 2, (Jan. 1971), p. 239–243.

153. A. Pelazzi, "Localized Distribution of Inclusions in Steel and Its Effect on the Confidence of Counting Data," *Rev. Met.*, V. 47, No. 12, (1950), p. 907–929.

154. T. Reti, "Application of Generalized Poisson Distribution to Describe the Numerical Distribution of Microscopic Particles, Especially of Oxide Inclusions," *Banyasz. Kohasz. Lapok*, V. 108, (Aug. 1975), p. 340–344.

155. I.D. Simpson and N. Standish, "Derivation of the Log-Normal Form of Both Section and Spatial Distributions of Particles," *Metallography*, V. 10, (1977), p. 149–159.

156. I.D. Simpson and N. Standish, "Determination of the Size-Frequency Distributions of Nonmetallic Inclusions in the As-Cast Metals," *Metallography*, V. 10, (1977), p. 433–449.

157. D.J. Widgery and J.F. Knott, "Method for Quantitative Study of Inclusions Taking Part in Ductile Fracture Processes," *Metal Science*, V. 12, (Jan. 1978), p. 8–11.

158. A.J. Birkle et al, "Analysis of Plain Strain Fracture in a Series of 0.45C–Ni–Cr–Mo Steels with Different Sulfur Contents," *Trans. ASM*, (1966), V. 59, p. 981–990.

159. D.E. Passoja and D.C. Hill, "On the Distribution of Energy in the Ductile Fracture of High Strength Steels," *Met. Trans.*, V. 5, (Aug. 1974), p. 1851–1854.

160. D.C. Hill and D. Passoja, "Understanding the Role of Inclusions and Microstructure in Ductile Fracture," *Welding J.*, V. 53, (Nov. 1974), p. 481s–485s.

161. N.E. Hannerz and J.F. Lowery, "Influence of Micro-Slag Distribution on MIG-MAG Weld Metal Impact Properties," *Met. Constr. Br. Weld. J.*, V. 7, (Jan. 1975), p. 21–25.

162. L.L. Chin, "A Model for Toughness Studies of Welds," *Welding J.*, V. 48, (July 1969), p. 290s–294s.

163. A.C. Steel, "The Effects of Sulphur and Phosphorus on the Toughness of Mild Steel Weld Metal," *Welding Res. Intl.*, V. 2, No. 3, (1972), p. 37–76.

164. D.A. Melford, "The Design and Use of Automatic Instruments for Cleanness Assessment," *ISI SR 112*, (1968), p. 14–23.

165. C. Fisher, "The Quantimet Image Analyzing Computer; *ISI SR 112*, (1968), p. 24—30.
166. J.A. Belk, "Inclusion Counting Methods: The Flying Spot Microscope," *ISI SR 77*, (1963), p. 25—29.
167. R.A. Bloom et al, "An Electronic Scanner — Computer for Determining the Non-Metallic Inclusion Content of Steel," *JISI*, V. 202, (Feb. 1964), p. 107—112.
168. J.L. Bayer et al, "Use of the AMEDA Microscope in Quantitative Microscopy," *ASTM STP 430*, (1968), p. 118—128.
169. D.M. Cottingham et al, "The TI—Hilger Optical Inclusion Counter," *ISI SR 112*, (1968), p. 31—39.
170. K.A. Ridal, "The Vickers Automatic Inclusion Classifier," *ISI SR 112*, (1968), p. 40—46.
171. K.A. Ridal and R. Cummins, "Automatic Equipment for Assessing the Cleanness of Steel," *ISI SR 134*, (1972), p. 248—254.
172. C. Fisher and M. Cole, "The Metals Research Image Analyzing Computer," *Microscope*, V. 16, No. 2, (April—July 1968), p. 81—94.
173. J. Gibbons et al, "A Description of the Quantimet 360 Inclusion and Grain Size Classifier," *Microscope*, V. 20, No. 1, (Jan.—April 1972), p. 1—20.
174. P.H. Lindon, "The Use of Image Analysis in Assessment of Growth and Separation of Deoxidation Products in Steel," *Microscope*, V. 16, No. 2, (April—July 1968), p. 137—150.
175. R. Roche, "The Use of the 'Quantimet' Microscope in the Micrographic Quantitative Determination of Combined Oxygen at Oxide Inclusions," *Microscope*, V. 16, No. 2, (April—July 1968), p. 151—161.
176. T.R. Allmand, "Proving Trials of the Quantimet Image Analyzing Computer in Metallography," *Microscope*, V. 16, No. 2, (April—July 1968), p. 163—170.
177. J.R. Blank, "Nonmetallic Inclusion Assessment by Image Analysis Techniques," *Microscope*, V. 16, No. 2, April—July 1968, p. 189—197.
178. T. Kelly, "Preliminary Evaluation of the Quantimet 360 Inclusion and Grain Size Classifier," *Microscope*, V. 20, No. 1, (Jan—April 1972), p. 21—35.
179. T.R. Allmand and D.S. Coleman, "Technical Problems in Assessing Non-metallic Inclusions in Steel for Quality Control," *Microscopy*, V. 20, No. 1, (Jan—April 1972), p. 57—81.
180. R. Widdowson, "An Evaluation of the Metals Research Prototype Quantitative Television Microscope for Assessment of the Inclusion Content in Steel," BISRA Report SM/B/4/65, (1965).
181. R. Ruddlestone et al, "Assessment of Steel Cleanness Using Quantimet 360," *Metals Tech.*, (Sept. 1960), p. 422—432.
182. J.J. Gautier et al, "First Applications of Texture Analysis to Metallography," *ISI SR 134*, (1972), p. 255—263.
183. T.R. Allmand and J.R. Blank, "An Evaluation of the Quantimet Image Analyzing Computer for Assessing Non-metallic Inclusions and Other Microscopical Features in Metals and Alloys," *ISI SR 112*, (1968), p. 47—71.
184. S. Johansson, "Rating Nonmetallic Inclusions by Image Analysis," *ASTM STP 575*, (1975), p. 163—177.
185. T.R. Allmand, "Results Obtained with the Fully Automatic Metals Research Inclusion Counter," GKN Group Tech Centre Report No. 872, (Nov. 1966).
186. T.R. Allmand, "Inclusion Counting. An Evaluation of the TI/Hilger and Watts' Automatic Inclusion Counter L148," GKN Research Report No. 957, (Sept. 1967).
187. D.A. Melford, "Automatic Cleanness Assessment of Steel," *ISI SR 134*, (1972), p. 229—234.
188. H. Martensson, "Report of the Jernkontoret Automatic Inclusion Assessment Committee," *ISI SR 134*, (1972), p. 235—240.

189. S. Johansson, "Determination of Non-metallic Inclusions in Steel by Image Analysis," *Scand. J. Met.,* V. 2, (1973), p. 24–28.

190. R.A. Rege et al, "Microcleanliness of Steel –– A New Quantitative TV Rating Method," *ASTM STP 480,* (1970), p. 249–272.

191. S. Johansson, "Image Analysis in Quality Control of Steel," *Microscope,* V. 20, No. 1, (Jan–April 1972), p. 83–90.

192. S. Johansson, "Design Criteria in the Quality Control of Steel with Respect to Nonmetallic Inclusions," *4th Intl. Congress for Stereology, NBS SP 431,* (1976), p. 261–264.

193. D. Jeulin and J. Serra, "Automatic Recognition of Non-metallic Stringers in Steels," *4th Intl. Congress for Stereology, NBS SP 431,* (1976), p. 265–268.

194. H.J. Kostler, "Validity of Measurements of Manganese Sulphide Shape Using Quantitative Micrographic Analysis," *Arch. Eisenhutt.,* V. 45, No. 11, (1974), p. 813–816 (BISI 13016).

195. C.H. Engstrom and K.O. Karlstrand, "The Uddeholm Method for the Assessment of Non-metallic Inclusions in Steel, Based on Quantitative Image Analysis," *Scand. J. Met.,* V. 2, (1973), p. 105–108.

196. H. Brandeis et al, "The Applicability of the Quantitative Television Microscope (QTM) to Purity Assessment of Steels," *Prakt. Met.,* V. 6, No. 12, (1969), p. 728–741.

197. M. Frohlke, "Determination of Characteristic Values for Manganese Sulphides in Free-Cutting Steels by Means of Quantitative Image Analysis," *Microscope,* V. 19, (1971), p. 403–414.

198. A.G. Franklin, "The Sampling Problem and the Importance of Inclusion Size Distribution," *ISI SR 134,* (1972), p. 241–247.

199. A.G. Franklin, Comparison Between A Quantitative Microscope and Chemical Methods for Assessment of Non-metallic Inclusions," *JISI,* V. 207, (Feb. 1969), p. 181–186.

200. G.A. Moore, "Automated Measurements for the Determination of the Statistical Nature of the Distribution of Inclusions in Steel," *4th Intl. Cong. for Stereology, NBS SP 431,* (1976), p. 197–198.

201. T.R. Allmand and D.S. Coleman, "Techniques for the Determination of Non-metallic Inclusions in Steel," *Metals and Materials,* (Oct. 1971), p. 325–335.

202. S. Johansson, "The Sandvik Method for the Assessment of Non-metallic Inclusions in Steel with the Quantimet TV Microscope," *Jern. Ann.,* V. 154, No. 9, (1970), p. 423–426.

203. T. Hersant and D. Jeulin, "Sampling in Quantitative Image Analysis," *CDS Circ.,* (1977), No. 3, p. 733–755 (BISI 16624).

204. D. Jeulin and J. Serra, "To Recognize Inclusions: Charts or Texture Analyzer?" *CDS Circ.,* (1976), No. 5, p. 1165–1177 (BISI 14972).

205. R. Roche, "Quantitative Microscopic Metallography. A Critical Study of the Quantimet Quantitative Microscope. Examples of Its Use," *Meteau–Corrosion Ind.,* V. 43, No. 510, (Feb. 1968), p. 49–79.

206. B.L. Dobrotsvetov, "Using the Quantimet Automatic TV Microscope in Determinations of Oxygen Content in Blister Copper," *Ind. Lab.,* V. 39, (Nov. 1973), p. 1759–1762.

207. B.L. Dobrotsvetov, "Optimization of Parameters of the Methods of Determining Oxygen in Black Copper on an Automatic Model of the Television Microscope 'Quantimet'," *Ind. Lab.,* V. 42, (Feb. 1976), p. 232–235.

208. W. Jagiello-Puczka and W. Klimecki, "Inhomogeniety of Non-metallic Inclusions in Rail Steels. I. Quantitative Microscope," *Chemia Analityczna,* V. 20, No. 4, p. 791–798.

209. S. Malm, "On the Speed and Accuracy of Modern Image Analyzers," *ISS Newsletter '77 in Stereology,* (Dec. 1977), p. 55–80.

210. W. Bartholome et al, "Comparison of Results of Stereological Measurements with the Chemical Analysis of Fe-Mn-S Alloys with the Leitz Classimat," *Leitz Sci. and Tech. Information,* V. 6, No. 4, (June 1975), p. 127–132.

211. D. Radtke and D. Schreiber, "Relationship Between Sulphide Formation and Machinability in Free-Cutting Steels," *Steel Times,* V. 193, No. 5118, (Aug. 1977), p. 246–259.

212. S. Johansson and S. Malm, "A Study of Micro Inclusions in Stainless Steels Made by the CLU-Converter Process," *Scand. J. Met.,* V. 7, No. 1, (1978), p. 18–25.

213. E. Grethen, "The Influence of a Chromic Acid Etch on the Results of the Determination of Sulphides by the Quantimet," *Prakt. Met.,* V. 6, No. 11, (Nov. 1969), p. 672–675.

214. E. Beraha, "Metallographic Reagent for Identification of Sulphide Inclusions," *JISI,* V. 202, (Aug. 1964), p. 696–698.

215. E. Beraha, "Metallographic Reagent for the Identification of Sulphide Inclusions in Cast Iron and Steel and for the Differentiation of Phosphide and Cementite in Cast Iron," *Prakt. Met.,* V. 6, No. 9, (1969), p. 565–568.

216. A. Salak and A. Sol'om, "Determination of Sulphide Inclusions in Sintered Steel with Beraha's Reagent," *Pokroky Prask. Met.,* No. 2, (1965), p. 32–37.

217. A. Salak and A. Sol'om, "Metallographic Reagent for Exposing Sulfur Inclusions by Beraha's Method," *Hutnicke Listy,* V. 20, No. 10, (1965), p. 737–738.

218. R. Desbrandes et al, "Metallographic Identification of Sulfide Inclusions In Steels," *Rev. de l'Inst. Francais du Petrole et Annales des Comb. Liq.,* No. 3, (1965), p. 598–610.

219. A.E. Gerds and C.W. Melton, "New Etch Spots Leaded Steels," *Iron Age,* V. 178, (Aug. 30, 1956), p. 86–87.

220. K.E. Volk, "The Metallographic Proof of Pb in Steel," *Arch. Eisenhutt.,* V. 16, (1942), p. 81–84.

221a. Anon, "Occurrence of Lead in Lead-Bearing Steels," *The Iron Age,* V. 156, (September 20, 1945), p. 83, 162, 163.

221b. Anon, "Microscopy of Lead-Bearing Steels and Irons," *The Iron Age,* V. 156, (July 12, 1945), p. 49, 130.

222. W. Pepperhoff, "Structure-Developments by Means of Interference Vapour Deposited Films," *Arch. Eisenhutt.,* V. 32, (April 1961), p. 269–273 (BISI 3111).

223. W. Pepperhoff, "Quantitative Vertical Illumination Microscopy with the Help of Evaporated Interference Layers," *Arch. Eisenhutt.,* V. 36, (Dec. 1965), p. 931–950.

224. H.E. Buhler et al, "Possibilities of the Quantitative Description of Oxide Phases by Means of Evolving the Structure Without Etching According to W. Pepperhoff," *Arch. Eisenhutt.,* V. 41, (May 1970), p. 405–411.

225. S.D. Wicksell, "The Corpuscle Problem. A Mathematical Study of a Biometric Problem," *Biometrika,* V. 17, (1925), p. 84–99.

226. S.D. Wicksell, "The Corpuscle Problem. Second Memoir. Case of Ellipsoidal Corpuscles," *Biometrika,* V. 18, (1926), p. 151–172.

227. R.T. DeHoff, "The Determination of the Size Distribution of Ellipsoidal Particles from Measurements made on Random Plant Sections," *Trans. AIME,* V. 224, (1962), p. 474–477.

228. R.T. DeHoff, "The Estimation of Particle Size Distributions from Simple Counting Measurements made on Random Plant Sections," *Trans. AIME,* V. 233, (1965), p. 25–29.

229. E. Ekelund and T. Werlefors, "A System for the Quantitative Characterization of Microstructures by Combined Image Analysis and X-Ray Discrimination in the Scanning Electron Microscope," *Scanning Electron Microscopy,* (1976), 1, p. 417–424.

230. T. Werlefors and S. Ekelund, "Automatic Multiparameter Characterization of Non-metallic Inclusions," *Scand. J. Met.,* V. 7, No. 2, (1978), p. 60–70.
231. R.T. DeHoff, "Quantitative Metallography," *Techniques for the Direct Observation of Structures and Imperfections,* J. Wiley & Sons, NY, (1968), p. 221–253.
232. R.T. DeHoff, "Quantitative Microstructural Analysis," *ASTM STP 430*, (1968), p. 63–95.

# APPLICATIONS OF AUTOMATIC SPECIMEN–PREPARATION TECHNIQUES IN THE METALLOGRAPHY OF STEEL

W.D. Forgeng, Jr.* and A.G. Lee, Sr.*

## INTRODUCTION

The advent of automatic image analysis (AIA) for the measurement of micro-structural features in steel has placed two very demanding requirements upon methods for preparing metallographic specimens. First, the use of AIA requires that specimens be prepared uniformly free of relief, pitting, dragging, scratches, and other artifacts that might be detected by the instrument and erroneously measured. Second, because most AIA instruments are capable of measuring a field in a few seconds or less, the preparation method should be one that enables the metallographer to produce polished specimens at a rate sufficiently rapid to keep the expensive AIA instrument in use most of the time. These two requirements — high quality of specimen surface and speed — are often incompatible. In fact, we maintain that both requirements cannot be satisfied by using hand methods of preparation and that automatic specimen preparation is a necessity for AIA. Furthermore, the use of semiautomated metallographs and the now almost universal acceptance of Polaroid$^{(R)}$ photomicrography have increased the speed with which a metallographer can and should be capable of producing micrographs of excellent quality, and it is imperative that methods of specimen preparation keep pace with this more traditional phase of metallography.

U.S. Steel Research has developed an automatic method for preparing metallographic specimens of excellent quality for AIA and for metallographic examination and photomicrography. Commercially available automatic grinding and polishing equipment and expendable supplies are utilized. To achieve satisfactory results, it was necessary to make modifications, some of which are significant, to the techniques originally recommended by the equipment manufacturer and the abrasive suppliers. In this paper, the USS Research techniques and the results of their use are described.

## SPECIMEN–PREPARATION TECHNIQUES

### Initial Preparation

Specimen size may be selected to allow mounting in 1-, 1¼-, or 1½-inch-diameter mounts, or the specimens may be left unmounted if they are larger than about 1 by 1 inch up to about 1 by 3 inches in size and up to 3/4 inch thick. The specimens should be machine surface ground to produce a finish comparable to a number 80 grit; however, in the event that a surface grinder is not available, belt grinding or other methods capable of producing such a finish may be used. The pregrinding

*U.S. Steel Corporation, Monroeville, Pennsylvania USA.

of the specimens, whether by machine or by hand, should be accomplished so as to produce as flat a surface as possible. If the specimens are to be mounted, all sharp corners and burrs that can induce cracks to form in the mounting medium should be removed before mounting.

Mounting

When preparing mounted specimens, we normally use relatively inexpensive Bakelite metallographic mounts; however, when edge preservation is important, a small amount of filled epoxy resin is used to cover the specimen before the mounting press is filled with the Bakelite compound to complete the mounting. For speed and uniformity of mounting, it is recommended that pneumatic mounting presses operating on 80-psi air pressure for 1- and 1¼-inch mounts or 90- to 100-psi air pressure for 1½-inch mounts and molding temperatures of 280 to 300°F be used. Slightly lower pressures and curing times about 50 to 100 percent longer than those recommended by the manufacturer will result in harder mounts, which are useful especially when the epoxy-resin edge-preservation mounting techniques previously described are used. Following ejection of the mount from the press, it is only necessary to remove the excess resin flash from the mount before proceeding to automatic grinding and polishing. Belt grinding of the specimen samples after mounting is *not* recommended inasmuch as mold-platen surfaces of mounting presses are usually parallel and ensure a flat mount prior to automatic surface preparation. If the specimen height is less than 3/8 inch, two specimens may be mounted "back-to-back" at the same time, and this double mount can then be separated with a cutoff wheel as shown in Fig. 1; this procedure effectively doubles mounting-press capacity.

Fig. 1.   Method of doubling mounting-press capacity by producing "back-to-back" Bakelite metallographic mounts.

Mounting of the Specimens in Appropriate Holders

Up to six specimens are placed in an appropriate holder, (Fig. 2), and the holder is placed in a loading fixture that maintains the flatness of the specimens. The
specimens are locked in place with setscrews, and the holder is removed from the
loading fixture for fine grinding and final polishing of the mounted specimens. Unmounted specimens are coarse-machine-ground in the holder before fine grinding and
polishing.

Grinding

The holder is driven in a direction opposite that of the grinding wheel, which
is covered with a silicon-carbide paper disc of correct grit size. Water is used as a
lubricant. Grinding time may be from 1 to 3 minutes for each abrasive disc. De-
pending on total specimen-surface area, from one to three holders may be processed
with each disc. After each grinding, the holder is thoroughly washed in a liquid
soap-water mixture. Washing in this soap-water mixture is recommended after each
grinding without exception through the final (600 grit) grinding step. For steels
that are particularly prone to pitting and staining, such as low-carbon sheet steels or
electrical steels, an alcohol rinse, rather than the soap-water mixture, is recommended
after the final grinding step.

Polishing

A major difference between our polishing procedure and that recommended by
most manufacturers of automatic polishing equipment is the elimination of 9-micron-
or 6-micron-diamond polishing, which we have found to be an unnecessary step when
the 3-micron- and 1-micron-diamond polishing are done with a low-nap rayon cloth
and an alcohol-glycol lubricant. Table I is intended only as a guide covering the
grinding and polishing procedures for well-retained nonmetallic inclusions and a matrix free of disturbed metal and other artifacts.

## EXAMINATION OF POLISHED SPECIMENS

Following final polishing with the 1-micron-diamond cloth, the holder is thoroughly washed in an alcohol bath, dried with a blast of hot air, and then placed in-
tact under the microscope. Each specimen is examined for inclusion retention and
absence of scratches while in the holder. If minute scratches persist, the holder is
polished again with the 1-micron-diamond cloth, using little or no pressure, washed,
and reexamined. When the specimens are acceptable, they are removed from the holder and returned to the alcohol bath for a final washing. After they are washed,
the specimens are lightly rubbed with a dry cotton ball while they are dried with a
dry cotton ball while they are dried with a hot-air blast. Specimens not satisfactory
for further work must be reground and should not be removed from the holder because reinsertion does not maintain flatness. Satisfactorily polished specimens await-
ing analysis should be stored in a desiccator until they can be examined. When long
storage times are required, the specimens can be sprayed with clear acrylic lacquer,
which can be later removed with acetone followed by an absolute-methyl-alcohol
rinse.

For AIA, specimens are placed directly on the appropriate quantitative measuring instrument without additional hand preparation. Likewise, if etched specimens
are being routinely examined up to magnifications of X1000 or are being photographed
up to magnifications of about X500, the specimen surface may be etched directly

TABLE I

Final grinding and polishing procedures

| Stage | Abrasive and Size | Wheel Covering | Lubricant | System "B" | | | System "S" | | |
|---|---|---|---|---|---|---|---|---|---|
| | | | | Wheel Speed, rpm | Pressure Adjustment, lb | Sequence Time, min | Wheel Speed, rpm | Pressure Adjustment, lb | Sequence Time, min |
| Grinding | 240-grit SiC | Self-adhesive paper | Water | 163 | 60 | 2-3 | 150 | 30 | 1 |
| Grinding | 320-grit SiC | Self-adhesive paper | Water | 163 | 60 | 2-3 | 150 | 30 | 1 |
| Grinding | 400-grit SiC | Self-adhesive paper | Water | 163 | 60 | 2-3 | 150 | 30 | 1 |
| Grinding | 600-grit SiC | Self-adhesive paper | Water | 163 | 60 | 2-3 | 150 | 30 | 1 |
| Preliminary polish | 3-micron diamond paste* | Self-adhesive synthetic cloth (rayon) | Diamond-paste extender** | 246 | 60 | 3 | 150 | 60 | 3 |
| Final polish | 1-micron diamond paste* | Self-adhesive synthetic cloth (rayon) | Diamond-paste extender** | 246 | 60 | 3 | 150 | 60 | 1-1/2 |

* Recharging of cloth may be accomplished with diamond-powder aerosol.

** Formula (to make 1 liter)
800 ml methyl alcohol
100 ml ethylene glycol
100 ml $H_2O$

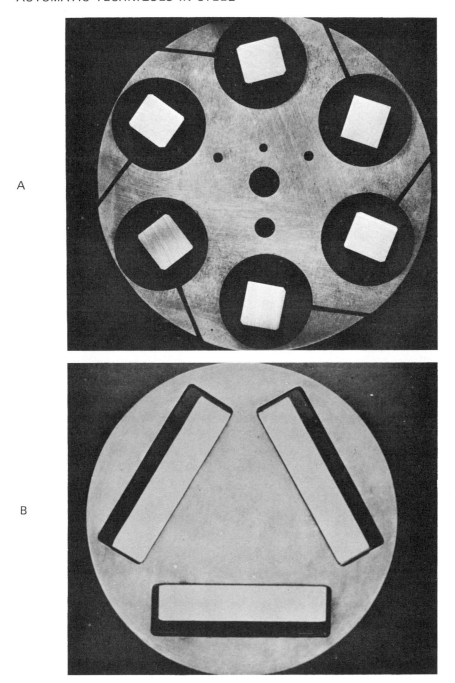

Fig. 2. Metallographic specimens in holders after rough grinding and before final grinding and polishing. A. Holder containing 6 mounted specimens ready for automatic grinding and polishing on 8-inch-diameter wheels. B. Holder containing 3 unmounted specimens ready for automatic grinding and polishing on 12-inch-diameter wheels.

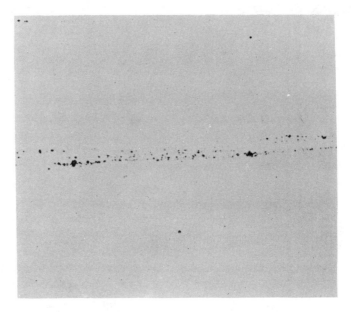

A

B

Fig. 3.   Alumina inclusions in a Si-Al-killed plate-steel specimen polished with U.S. Steel automatic procedure.   Longitudinal section.   A.   X100.   B.   X200.

A

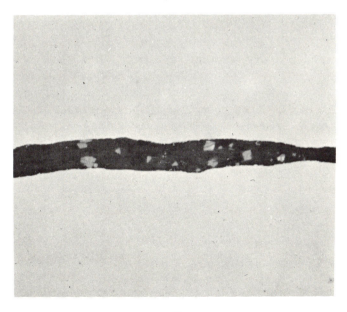

B

Fig. 4. Silicate inclusions in a Si-killed billet-steel specimen polished with U.S. Steel automatic procedure. Longitudinal section. A. X75. B. X400.

A

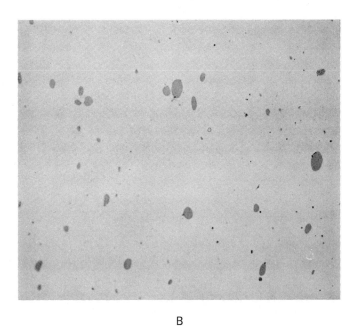

B

Fig. 5. Sulfide inclusions in specimens polished with U.S. Steel automatic procedure. A. Resulfurized stainless steel. Longitudinal section. X100. B. Resulfurized carbon steel. Transverse section. X400.

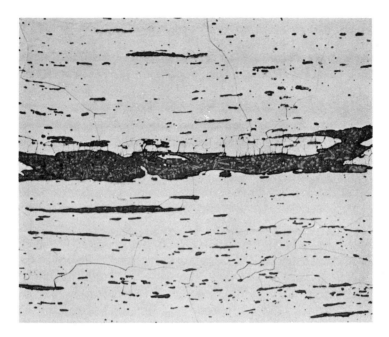

Fig. 6. Slag inclusions in a specimen taken from a 125-year-old wrought-iron beam and polished with U.S. Steel automatic procedure. Nital etch. X75.

## TABLE II

Effect of preparation method on automatic-image-analysis inclusion measurements for a silicon-killed carbon-steel billet sample

| Preparation | AIA Inclusion Measurement | | |
| | Area Percent | Projected Length, $mm/mm^2$ | Count per $mm^2$ |
| --- | --- | --- | --- |
| Automatic #1 | 0.35 ± 0.04* | 0.48 ± 0.05 | 37 ± 3 |
| Hand | 0.47 ± 0.05 | 0.66 ± 0.06 | 44 ± 3 |
| Automatic #2 | 0.31 ± 0.04 | 0.42 ± 0.06 | 27 ± 3 |

* 95 Percent confidence limits.

Note: 20 $mm^2$ of specimen-surface area rated with X10 0.2 N.A. objective.

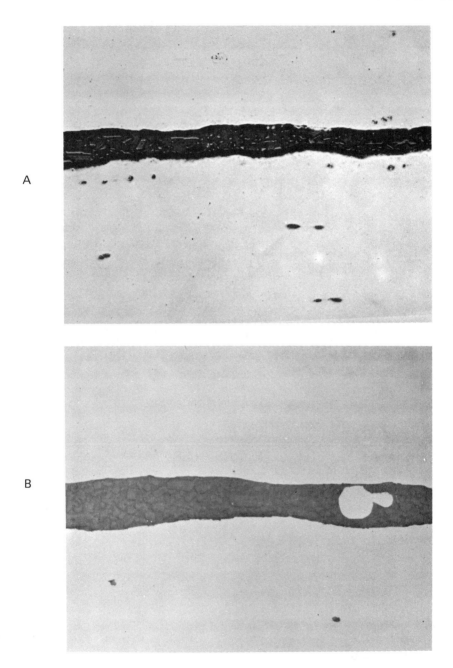

Fig. 7. Comparison of hand and automatic specimen-preparation techniques for the polishing of a steel specimen containing large silicate inclusions. X200. A. Prepared at another laboratory by using hand method. B. Reprepared at U.S. Steel by using automatic method.

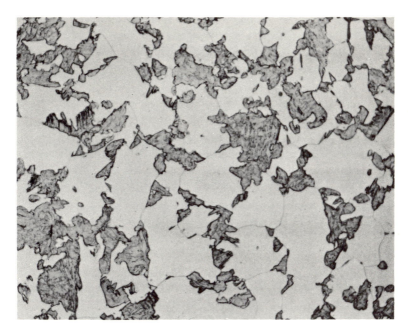

Fig. 8. Dual-phase (ferrite-martensite) sheet steel. Picral etch immediately following automatic polish on 1-micron diamond. X500.

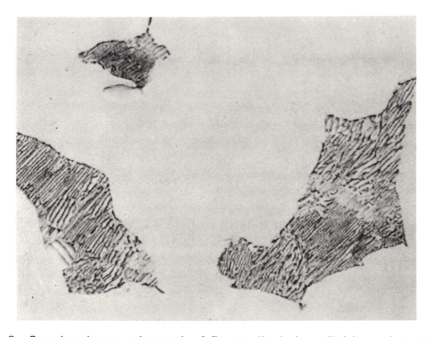

Fig. 9. Scanning electron micrograph of fine pearlite in hot-rolled low-carbon steel. Final polish by hand with 0.03-micron alumina after automatic polish. Picral etch. X3000.

without concern for "disturbed metal," which is minimized by using diamond as a final polishing medium. High-magnification light microscopy and transmission or scanning electron microscopy should be preceded by one etch-polish-etch cycle by hand to remove the very small amount of disturbed metal that exists and to remove fine scratches caused by the 1-micron diamond (final hand polishing is typically accomplished with 0.03-micron alumina).

## RESULTS OF THE METHOD

Photomicrographs of oxide- and sulfide-type nonmetallic inclusions in steel specimens polished by using our method are shown in Figs. 3 to 6. The hard oxide and relatively hard silicate inclusions are well retained and are free of relief, and the softer sulfide inclusions are free from pitting and dragging. Figure 7 shows a large, duplex silicate inclusion in a steel specimen that was polished by hand and then repolished in accordance with the automatic procedure we have described. Pullout and grooving along the edges of the inclusion, which resulted from hand polishing and could have been misinterpreted as a lamination in the steel, have been eliminated with the proper automatic polishing procedure.

Quantitative-television-microscopy measurements of inclusion parameters for a specimen prepared first with our automatic method, then with careful hand repolishing, and again with automatic repolishing are given in Table II. The values for inclusion area, length, and number were significantly higher on the hand-polished sur-face than on either of the automatically polished surfaces.

Figures 8 and 9 illustrate the use of our automatic method for steel specimens that are normally difficult to prepare for examination in the etched condition. Figure 8 shows a dual-phase (ferrite-martensite) sheet steel, etched directly in picral after automatic polishing. The relatively hard martensite is well defined and the ferrite suffers no apparent degradation. Figure 9 is a scanning electron micrograph of very fine pearlite in a hot-rolled low-carbon steel. Only a minimum of hand repolishing was used between two etching steps to produce the excellent definition of the pearlite lamellae.

## SUMMARY

The use of automatic methods for preparing metallographic specimens provides the best quality section for the AIA measurement of nonmetallic inclusions in steel and is recommended for routine metallographic work as a time-saving procedure. Several changes are suggested in procedures recommended by the equipment and con-sumable-supply manufacturers to save time and produce optimum results. Specimens prepared with the methods recommended herein are free from relief, inclusion pull-out, and other major deficiencies that can cause significant errors in the measurement of inclusion content, size distribution, and morphology and in the interpretation of microstructures.

# CHARACTERIZATION OF METALWORKED SURFACES

Roger N. Wright*

Surface characterization is a steadily growing element of metalworking quality control. Part of this is due to enhanced instrument capability, such as the great depth of field afforded by the scanning electron microscope (SEM). Some of the interest is related to increased awareness of the role of the surface character in such field service problems as fatigue, corrosion, stress corrosion cracking, and so on. Beyond this, increased focus on "net shape processes" has led to consideration of "finished" surface qualities far different from those developed by machining, grinding, and other surface conditioning operations.

In this paper, some aspects of the basic nature and interpretation of the metalworked surface will be reviewed for the metallographer. Little emphasis will be placed on metallographic technique, per se. Rather, the emphasis will be placed primarily on the origin of surface detail. The interested reader is referred to previous discussions by the author, as well[1,2].

In broad terms the nature of the metalworked surface is encompassed by the following categories:
   a)  topography
   b)  crack incidence
   c)  residual stress state
   d)  chemistry
   e)  microstructure and substructure
This paper will focus on items (a) and (b). Item (c) is, at this time, an extraordinary elusive measurement, especially in view of the wide local variations that may be presumed to exist. Definitive measurements of items (d) and (e) are well within reach for quality control laboratories equipped with facilities for Auger electron spectroscopy, transmission electron microscopy, electron beam microprobe analysis, and so forth. However, sample preparation time and instrument running time can be sufficiently lengthy to preclude *routine* quality control analysis in many cases. Even so, the interested reader may want to review the recent work of Doig and Flewitt[3] which deals quite succinctly with the measurement of near-surface substructure gradients, hardness gradients, and their probable origins in the case of simple tensile elongation.

Metalworked surface topography and cracking reflect the die or tool, the "workpiece" (the metal being worked), and the intervening conditions of friction and lubrication. While lubrication is an extremely complicated subject, mechanically as well as chemically, typical tool-workpiece interface conditions can be grouped into the

*Rensselaer Polytechnic Institute, Troy, New York 12181  USA.

following modes:
1)  sticking friction;
2)  boundary lubrication;
3)  thick film lubrication.

Sticking friction is simply that case where the workpiece sticks to the die. The implication is that there is no lubricant or that the lubrication has somehow broken down. Such conditions are *typical* in many hot working operations. On the other hand, they usually reflect catastrophic lubricant breakdown in cold forming operations. When the workpiece sticks to the tool, the metal moving past the tool must tear away from itself. The frictional shear drag is no longer the shear strength of a lubricant, but rather the shear strength of the metal itself. The value of the coefficient of friction may approach one-half, although values in the range of 0.25 are probably more realistic. The emerging surface is, in effect, a ductile fracture surface showing many torn areas. Graphic illustrations of such surface detail are shown in Figs. 1 and 2 for the respective cases of rod rolling and wire drawing.

Boundary lubrication is said to prevail when a very thin layer of lubricant separates the tool from the workpiece. This film is, however, so thin that the workpiece surface rigorously conforms to the profile dictated by the tooling surface. Thus, in the face of smooth tooling, a smooth, highly reflective workpiece surface is developed. Even so, the coefficient of friction can be rather high (0.10 to 0.15) in view of all the surface metal flow that is required for the surface smoothing. A boundary lubricated surface can be almost featureless in its microscopic appearance. However, a thin line often exists between boundary lubrication and lubricant breakdown which leads to surface sticking. Sometimes local variations in such frictional conditions can produce intense surface shearing and related shear fractures. An apparent example of such shearing behavior is seen in Fig. 3. Such local cracking does little to diminish the overall glossiness of the surface, and, in fact, many "good looking" metalworked surfaces contain insidious flaws of the type shown in Fig. 3.

Thick film lubrication refers to conditions where the lubricant film thickness far exceeds the size of the die surface detail. Thus, the workpiece surface is not ironed out to conform intimately to the die profile and a smooth surface almost never develops. The frictional shear drag is simply the shear strength of the lubricant and the coefficient of friction, if relevant, can be very low (0.01–0.05). Thick film behavior can be associated with solid soap lubricants or with thick liquid films. Thick liquid lubricant films are referred to as hydrodynamic lubrication. Under fully hydrodynamic or thick film lubrication conditions, a matte or microscopically lumpy surface develops which can reflect grain boundaries and even shear bands and slip lines. The rough surface merely reflects a lack of the surface constraint that would have been imposed with boundary lubrication. Differential surface flow can be great enough even to lead to surface fractures. A surface generated under conditions of thick film lubrication is shown in Fig. 4.

Often a condition exists which is intermediate between thick film and boundary lubrication. That is, local regions of thick film develop. Such regions are referred to as "hydrodynamic pockets" and may punch rough depressions into a relatively smooth, boundary lubricated surface. An apparent example of such a surface is shown in Fig. 5.

Roughened or cracked metalworked surface quality can be progressively smoothed or lapped over when boundary lubrication is developed in subsequent aworking. While such "healing" may or may not have a restoring effect on surface sensitive engineering properties, it can certainly complicate metallographic interpretation. Figure 6 shows a surface that has been drawn with largely boundary lubrication

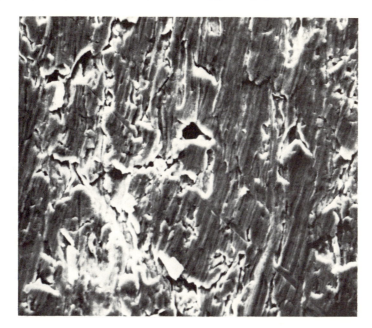

Fig. 1. SEM micrograph of the surface of hot rolled EC aluminum rod showing laps, slivers and torn areas indicative of sticking friction conditions. Rolling direction is nearly vertical. 400X. (Reduced 10% for reproduction.)

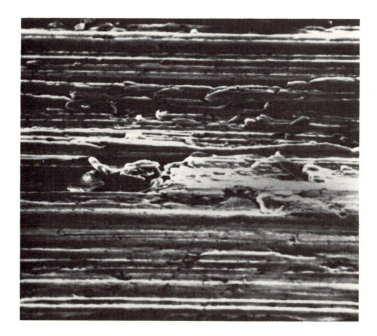

Fig. 2. SEM micrograph of the surface of cold drawn EC aluminum rod showing heavy longitudinal striations indicative of tearing during conditions of sticking friction. 400X. (Reduced 10% for reproduction.)

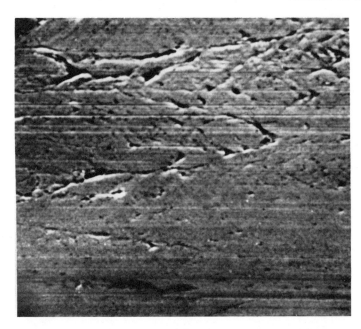

Fig. 3. SEM micrograph of the surface of EC aluminum wire showing local shear cracking in an otherwise smooth, boundary lubricated surface. A compounded mineral oil lubricant was used. Drawing axis is horizontal. 400X. (Reduced 10% for reproduction.)

Fig. 4. SEM micrograph of the surface of EC aluminum wire showing gross surface roughness typical of thick film lubrication. A calcium stearate lubricant was used. Drawing axis is horizontal. 400X. (Reduced 10% for reproduction.)

Fig. 5. SEM micrograph of the surface of EC aluminum wire, showing pockmarks probably caused by thick film lubricant pockets. A calcium stearate lubricant was used. Drawing axis is horizontal. 400X. (Reduced 10% for reproduction.)

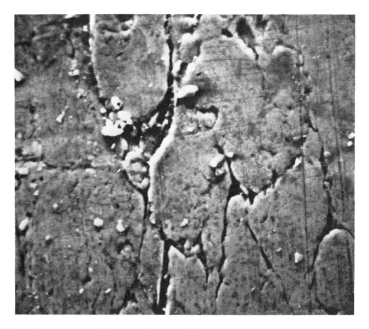

Fig. 6. SEM micrograph of the surface of EC aluminum wire, as shown from the condition of Fig. 1 with a compounded mineral oil lubricant. Drawing axis is vertical. 400X. (Reduced 10% for reproduction.)

from the condition shown in Fig. 1. Note how the torn surface has been smoothed. The surface may even appear smooth and bright to the naked eye. Even so, the crevices and laps that remain may be quite important to engineering performance.

Within the realm of boundary lubrication, there may be significant surface topography. First of all, the tool surface may not be entirely smooth. Figure 7 shows longitudinal striations representative of irregularities on a rod shaving die. Beyond this, smooth boundary lubricated surfaces are easily abraded upon contact with guides, sheaves, capstans, reels, or adjacent workpiece surfaces. An example of a smooth surface marred by such abrasion is shown in Fig. 8.

Much practical use can be made of surface quality characterizations in quality control. A case history involving the relation of electrical conductor surface flaws to insulation integrity has been summarized previously by the author[1]. It is important to realize that microscopic surface characterization may not be practically applied to all types of metalworking surface detail. In particular, it cannot be used to assess macroscopic defects and widely spaced, intermittent phenomena. On the other hand, continuous, or general microsurface detail can be quite successfully characterized. Since practical lubrication states are likely to persist over an extended length of time, the surface detail outlined above does, in fact, fit into this continuous general category. Samplings can be taken from coil ends, for example, with reasonable hope of obtaining surface quality comparable to that of the coil interior. Individual samples can be roughly scanned at 100X over areas the order of 25 mm$^2$. However, most of the detail described above can be best examined at somewhat higher magnification and representative micrographs should be taken in the 400X range.

Development of practical surface quality control procedures should generally adhere to the following steps:

i) Broadly review the product surface appearance and break surface detail into components representative of starting stock surface quality, lubrication practice, handling abrasion and so on. If possible assembly an atlas of representative micrographs showing various degrees of the surface detail components. It may even be useful to assign a numerical rating to levels of surface detail defined by the representative micrographs.

ii) Develop statistical relationships between product surface character and engineering performance properties such as fatigue response, stress corrosion cracking response, enamelability, and so forth.

iii) Once the most practical property-surface quality relationships have been identified, streamline the surface examination technique to a form compatible with straightforward evaluation in a manufacturing context. Scanning electron microscopy is nearly always the most useful surface evaluation tool and it is absolutely necessary for phase (i) outlined above. In some instances, scanning electron microscopes are readily available for manufacturing quality control. However, in many cases, reliance on light microscopes (and even eye loupes) is necessary. If *careful* comparison is made to the definitive scanning microscope images, light microscopy can be a practical tool for determining the intensities of *well established* flaw types.

The type work outlined can be performed in, and/or coordinated through, many research facilities, and it is hoped that the analysis of metalworked product surfaces will be the focus of much basic research and development. The most pressing needs for additional work are, however, outside the research laboratory. A major effort must be made to familiarize manufacturing personnel with proper use of

Fig. 7.  SEM micrograph of EC aluminum rod surface showing light striations from shaving die surface irregularities.  Rod pulling axis is horizontal.  400X. (Reduced 10% for reproduction.)

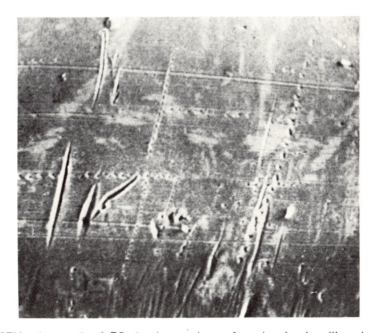

Fig. 8.  SEM micrograph of EC aluminum wire surface showing handling abrasions. Drawing axis is horizontal.  150X.  (Reduced 10% for reproduction.)

microscopy. This increased awareness would help promote much needed "shop floor" inspections and would also enhance communications among manufacturing, quality control, and research personnel. Other pressing needs are for a broadened data base of surface quality-property relationships and the expanded and enlightened use of surface quality criteria in product specifications and purchasing agreements.

It is a practical goal to make surface characterization as widely used and understood as microstructural evaluation. Surely surface quality is as important in many applications as grain size or inclusion morphology.

REFERENCES

1.  R.N. Wright, *Metal Progress,* 1978, Vol. 114, No. 3, p. 49.
2.  R.N. Wright and A.T. Male, *Trans. ASME, J. of Lubrication Tech.,* 1975, Vol. 97, Ser. F., p. 134.
3.  P. Doig and P.E.J. Flewitt, *Phil. Mag.,* 1977, Vol. 35, No. 4, p. 1063.

# SIMS/AES/XPS: COMBINED SURFACE ANALYSIS
# TECHNIQUES FOR QUALITY CONTROL

Robert M. Shemenski*

## INTRODUCTION

Some objectives for quality control specified by ASTM[1] are to discover physical causes of material behavior under particular conditions, to determine the origin of nonconformance with specified standards in order to eliminate assignable causes and to attain economic control of product quality, and to identify distributions of quality characteristics of materials which serve as a basis for setting economic standards of quality. Quality characteristics refer to quantitative information on measurable parameters of materials and manufactured products. Therefore, quality control techniques must collect quantitative data related to causes of observed phenomena. Surface quality is extremely important for optimum product performance so that both physical and chemical characterization of a surface are necessary dimensions of quality control. But in order to accurately characterize a surface, exactly what constitutes this surface must be defined.

Metallurgical engineers, metallographers, and others working in quality control routinely monitor macroscopic features affecting surface quality, such as quench cracks, improper heat-treatment or case hardening, forging or casting defects, inclusions, weld flows, improper machining or assembly, grinding cracks, chemical attack, corrosion, etc.[2,3]. In these cases, a "surface" may be several hundredths of an inch thick. For microelectronic components, catalysts, paints, adhesives, etc., where trace amounts of surface contaminants can severely degrade product quality, a "surface" is typically a few atomic layers thick[4,5].

Steel cords used in steel-belted radial tires are electrolytically coated with nominally 2000 Å of brass. This brass coating serves two functions: (1) it acts as a lubricant during the final wire drawing operation, and (2) it is the adhesive agent bonding steel cord reinforcement to rubber[6,7,8]. Historically, the tire industry utilized bulk copper composition and total brass thickness as a measure of the quality of brass coating. However, these standards might not be sufficient to insure good adhesion between steel and rubber. In fact, recent research has revealed that chemical composition of the outermost surface and partitioning of Cu/Zn throughout the total brass thickness are better benchmarks for predicting bond durability[9,10]. In the case of steel tire cord, quality control requires analysis of the outermost 10—30 A as well as successive atomic layers throughout the coating thickness. Consider also that there are fourteen major wire manufacturers worldwide supplying the tire industry, each having its unique processing, lubricants, surface treatments, etc. The rubber formulation of a given tire manufacturer must bond equally well to many microscopically different surfaces. This type of quality control requires

* The Goodyear Tire and Rubber Company, Akron , Ohio  USA.

109

specialized surface analysis techniques, viz., SIMS/AES/XPS[11,12,13,14]. Figure 1 gives the definitions of *surface analysis techniques* and *surface* used in this paper.

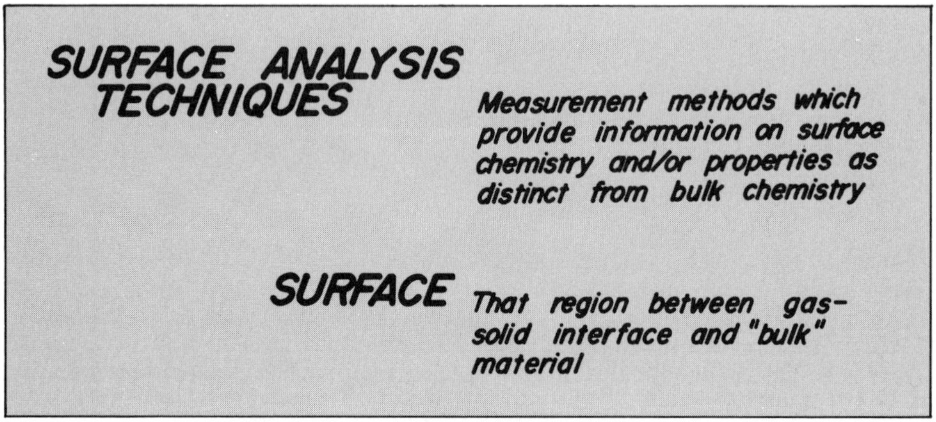

*SURFACE ANALYSIS TECHNIQUES* Measurement methods which provide information on surface chemistry and/or properties as distinct from bulk chemistry

*SURFACE* That region between gas-solid interface and "bulk" material

Fig. 1. Definitions used in this paper.

There are many spectroscopic techniques for surface analysis which operate basically on the same general principle[15,16]. A primary excitation beam is focused onto a surface to be analyzed causing excitation events to occur within the surface which emit a secondary beam. This secondary beam contains information about the chemical composition of the region bombarded by the primary beam. And since the excitation energy is low, the escape depth of emitted species is approximately 10–50 Å, thereby limiting chemical information obtained to near-surface layers. By using inert gas sputtering and continuously analyzing each newly created surface, chemical profile data throughout the thickness of a material may be obtained. Classification of spectroscopic techniques according to type of excitation and emission is shown in Fig. 2. The most common surface analysis techniques are outlined by dotted lines in Fig. 2. The acronyms for these surface analyses techniques are listed in Fig. 3.

The objectives of this paper are (1) to illustrate how surface analysis is used for quality control of brass surfaces on steel tire cord, and (2) to demonstrate how combined surface analytical techniques are more effective in isolating sources of surface quality contamination. The most common surface quality problems encountered on tire cord are variations in copper content and surface contamination, e.g., oxidation. For this study, these conditions were simulated by using solid brass coupons. Typical surface analysis data are presented to show how such quality deviations would be detected. This paper will not attempt to discuss in depth the mechanism of brass-to-rubber adhesion. The reader is referred to a comprehensive and critical review by Van Ooij[17]. Nor will this paper discuss the physics of each surface analysis technique. There are many fine papers on these subjects in the literature[18–30].

| EXCITATION ➡ | | | |
|---|---|---|---|
| | **PHOTONS** | **ELECTRONS** | **IONS** |
| **PHOTONS** | X RAY FLUORESCENCE INFRARED ABSORPTION RAMAN SCATTERING | ELECTRON MICROPROBE CATHODOLUMIN- ESCENCE APPEARANCE POTENTIAL | ION- INDUCED X RAYS |
| **ELECTRONS** | UV PHOTOELECTRONS *XPS ESCA* | LEED *AES SAM* | ION NEUTRALIZATION |
| **IONS** | *LAMMA* | ELECTRON- INDUCED DESORPTION | *SIMS ISS* |
| **NEUTRALS** | PHOTO DESORPTION | ELECTRON- INDUCED NEUTRAL DESORPTION | SPUTTERING |

*(EMISSION ⬇ along left side)*

Fig. 2. Classification of spectroscopic techniques according to type of excitation and emission as stated by Wehner [15].

*AES* → *Auger Electron Spectroscopy*

*SAM* → *Scanning Auger Microscopy*

*XPS* → *X ray Photoelectron Spectroscopy*

*ESCA* → *Electron Spectroscopy for Chemical Analysis*

*SIMS* → *Secondary Ion Mass Spectroscopy*

*ISS* → *Ion Scattering Spectroscopy*

*LAMMA* → *Laser Microprobe Mass Analysis*

Fig. 3. Explanation of acronyms used for surface analysis techniques.

EXPERIMENTAL

SAMPLES: Solid brass coupons are often used to study the effects of variations in brass surface chemistry on rubber adhesion[12,13,14,17]. This type of sample provides a large flat surface which is ideal for XPS analysis. Two soft, annealed
brass alloys supplied by Bridgeport Brass Company were used in this research, nominally 70/30 CDA–260 (69.45% Cu) and nominally 80/20 CDA–240 (79.60% Cu).
These alloys were received in sheet form 0.5 mm thick from which "T–shaped" sample coupons were cut. These coupons had an 8 mm x 8 mm test zone attached to
a 4 mm x 20 mm handle. The coupons were ground on 600-grit silicon carbide
paper and polished by using one-micro diamond paste. After polishing, samples
were ultrasonically cleaned in soapy distilled water and methanol, rinsed in double
distilled water, hot air dried, and stored in a desiccator. Several of these polished
coupons were then oxidized by heating to 275°C in air for two hours. Oxidized
coupons were stored in a desiccator until analyzed.

SURFACE ANALYSIS: As-polished and oxidized surfaces of both 70/30 and
80/20 brass coupons were analyzed by using SIMS, AES, SAM, and XPS techniques.
Figure 4 shows an overview of the complete integrated system.

This system was designed to permit simultaneous AES/SIMS plus XPS at one
position without moving the sample as well as simultaneous SAM/SIMS at another
position 180° away (see Fig. 5). Figure 6 shows a 12-position carrousel inside the
ultra-high vacuum chamber and the locations of differentially pumped ion gun, SAM
single-pass energy analyzer, and SIMS quadrupole. The specific operating parameters
for each surface analysis technique are listed below.

Fig. 4.   Overall view of surface analysis equipment showing SIMS electronics, vacuum
chamber, SAM/XPS/AES electronics, computer-plotter system.

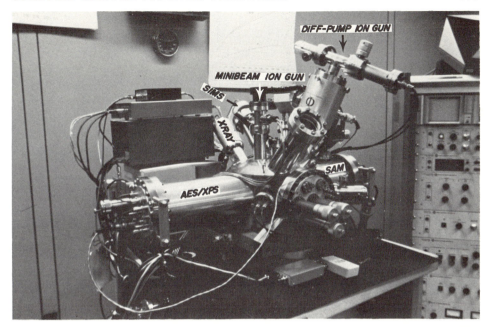

Fig. 5. Ultra-high vacuum chamber showing XPS/AES/SIMS analysis position on left (double-pass energy analyzer) and SAM/SIMS analyses position on right (single-pass energy analyzer).

Fig. 6. Sample carrousel inside vacuum system showing relative positions of differentially pumped ion gun, SAM energy analyzer, and SIMS quadruple.

SIMS:  A 3M Company Model 610 SIMS combined with a 3M Company MINI-BEAM ion gun system or a PHI Model 04–303 differential ion gun was used for secondary ion mass spectrometry.  Operating conditions for each run were:
  3M MINIBEAM ion gun:

  ion source gas —— $Ar^+_{40}$
  base pressure —— 5.5 x $10^{-5}$ torr (7.2 x $10^{-3}$ Pa)
  primary beam energy —— 2.5 keV
  sample distance —— 3.8 mm
  beam size —— medium (~500 $\mu$m)
  beam current density —— 60 $\mu$A/cm$^2$
  electron emission —— 15 mA
  electron energy —— 150 eV

  PHI differential ion gun:

  ion source gas —— $Ar^+_{40}$
  base pressure —— 1 x $10^{-7}$ torr (1.3 x $10^{-5}$ Pa)
  primary beam energy —— 3.5 keV
  sample distance —— 5 mm
  beam size —— 200 $\mu$m
  beam current density —— 318 $\mu$A/cm$^2$
  electron emission —— 25 mA

A UTI Model 100C quadrupole was used for mass analysis of the energy-filtered secondary ions (either positive or negative).  This quadrupole can also be used for residual gas analysis.  A Tracor-Northern Model NS–570 data processor/signal averager was used for data collection, storage, and arithmetic processing, viz., smoothing, normalization, background subtraction, etc.  These data were then plotted out on hard copy or digitally recorded on magnetic tape by using a Tracor-Northern Model NS–408C magnetic tape control and drive system.  The data recorded on magnetic tape are processed, analyzed, and plotted by using an IBM 370/168 computer system.

SAM:  A PHI Model 545 scanning Auger microprobe was used for surface analyses requiring high spatial resolution.  This technique was ideally suited for surface analysis of a single steel filament which would have a diameter in the range 175 $\mu$m–250$\mu$m.  Operating parameters used were:

  primary beam energy —— 3 keV
  electron beam current —— 10$\mu$A
  beam diameter —— 30$\mu$m
  base pressure —— 1 x $10^{-9}$ torr (1.3 x $10^{-7}$ Pa)

SAM data were collected and processed by using a PHI multiple-technique analytical computer system (MACS).

AES/XPS:  A PHI Model 548 ESCA/Auger electron spectrometer was used for AES and XPS analyses.  These are relatively large beam size techniques and so data obtained are more nearly average values over some finite surface area rather than microscopic spots as analyzed by SAM.  Operating conditions for each of these techniques were:
  AES:

  primary beam energy —— 3 keV
  electron beam current —— 10$\mu$A

beam diameter —— ~100$\mu$m
base pressure —— 1 x 10$^{-9}$ torr (1.3 x 10$^{-7}$ Pa)
XPS:

X-ray source —— MgK$a$ (1253.6 eV)
primary beam energy —— 10 keV
emission current —— 40 mA
beam diameter —— 3 mm

Data from both of these techniques were collected and processed by using a PHI multiple-technique analytical computer system. The XPS was calibrated by fixing the Au 4f 7/2 peak at 83.6 eV by adjusting the spectrometer work function used by the computer.

The energy distribution of secondary electrons produced by both AES and XPS is given by N(E). AES data were electronically differentiated to remove background due to backscattered primary electrons and inelastically scattered Auger electrons. Therefore, AES data were represented by plotting d[N(E)]/dE versus kinetic energy of emitted Auger electrons. The XPS analyzer was operated in a retarding mode giving an energy distribution function with intensity proportional to 1/E. XPS data were then given by plotting N(E)/E versus binding energy of photoelectrons.

MACS: The PHI multiple-technique analytical computer system provided indirect control over SAM, AES, and XPS, as well as providing complete data acquisition, data storage, multiplexing, signal averaging, and mathematical data processing. The MACS system consists of a PDP—1104 minicomputer with floppy disk storage, a keyboard input device, a high-speed digital plotter, and a PHI Model 20—137 computer interface.

PROCEDURE

The routine procedure for analyzing brass surfaces was to first run an XPS broad survey (50—1050 eV binding energy). After studying these broad-survey spectra, narrow range surveys over a single peak, e.g., ZnLMM at 261 eV, or a series of peaks, e.g., Cu 2p + Zn 2p at 930—1050 eV, were run. After finishing XPS analyses, AES or SAM broad surveys (50—1050 eV kinetic energy) were run on the as-received surface. Narrow-range Auger surveys were run where required. Next, in-depth profile Auger analyses through the brass surface were run by setting the Auger spectrometer to automatically count peak-to-peak heights of all elements of interest (up to 20 maximum) while continuously sputtering the surface. This is called multiplexing or profiling. While an Auger profile analysis was being run, the SIMS system was used to simultaneously analyze the secondary ions sputtered from the surface. The SIMS could be programmed to run broad surveys, either 0—100 amu or 100—200 amu, or it was multiplexed over a small mass range, e.g., 60—70 amu which contains all the copper and zinc isotopes. The ion gun was always rastered over an area much larger than the Auger or XPS beam diameter to preclude edge effects. After completion of a preselected sputter time, a final Auger broad survey and XPS broad and narrow surveys were run. All of the XPS and Auger data were permanently stored on floppy disks by MACS and all the SIMS data were stored on magnetic tape.

DISCUSSION

SIMS: SIMS analyses (broad surveys) of as-polished versus oxidized brass surfaces and 69.45% Cu versus 79.60% Cu coupons are compared in Figs. 7 and 8 for

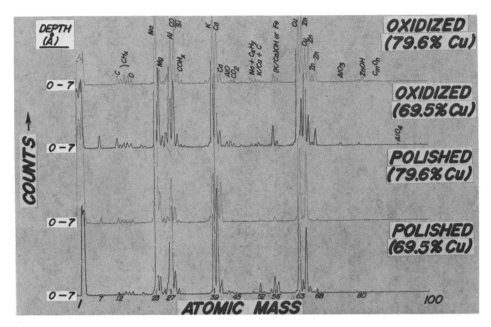

Fig. 7.  SIMS survey analyses from 0—100 AMU of as-polished and oxidized brass surfaces.

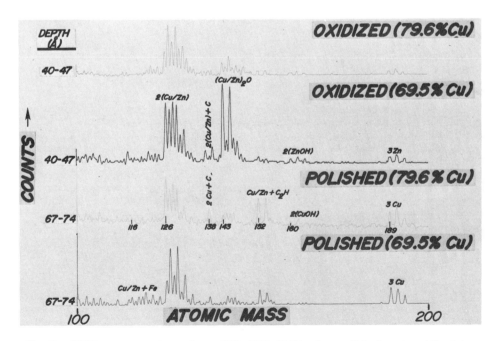

Fig. 8.  SIMS survey analyses from 100—200 AMU of as-polished and oxidized brass surfaces.

0—100 amu range and 100—200 amu range, respectively.  Note first that all spectra in Fig. 7 show very large peaks for sodium ($Na_{23}$), aluminum ($Al_{27}$), potassium ($K_{39}$), and calcium ($Ca_{40,41}$).  These are common contaminants which exist as trace amounts on most surfaces.  However, large peaks appear at these masses because sensitivities for such elements as Li, Na, and K are about 10X those for Mg, Al, and Ca, which are again 10X those for Fe, Cu, and Zn[31,32,33].  This analytical sensitivity is a function of such parameters as elemental sputtering yield ($Y_M$)[15,33,35], ionization ratio ($T_{M\pm}$)[35], ion yield ($K_M$)[35], and relative ionization coefficient ($P_M$)[35] as defined below:

$$Y_M = \frac{n(M)}{N_p} \tag{1}$$

$n(M)$ = total number of emitted atoms of element M
$N_p$ = number of incident primary ions

$$T_{M\pm} = \frac{n(M\pm)}{N_p} \tag{2}$$

$n(M\pm)$ = number of secondary ions of element M emitted

$$K_M = Y_M \cdot T_{M\pm} \tag{3}$$

$$P_M = \frac{K_M^x}{K_M^o} \tag{4}$$

$K_M^x$ = ion yield of solute element M in alloy x

$K_M^o$ = ion yield of pure element M

It is the influence of matrix effects and non-linear concentration dependence on the parameters given by Equations 1—4 that make quantification of SIMS particularly difficult[35].

As-polished surfaces for both 69.45% Cu and 79.60% Cu coupons were chemically similar except for copper level as seen by comparing $Cu_{63}/Zn_{64}$ ratios.  This would be expected since polishing gave an essentially "clean" surface.  Spectra from oxidized surfaces showed generally higher intensities than polished surfaces allowing identification of more specie on oxidized surfaces.  This occurred because an oxidized surface  generally has greater ion yield[36,37,38].  Although for as-polished surfaces, 80/20 brass coupons showed higher copper at the surface, upon oxidation at $275°C$ for 2 hours, a high level of zinc was found at the surface of 80/20 brass coupons.  These differences are even more apparent in the 100—200 spectra shown in Fig. 8.  As-polished surfaces were indicative of bulk chemistry, i.e., there was more copper than zinc at the surface and copper level at the surface of 80/20 brass coupons was higher than for 70/30 brass coupons.  On the other hand, oxidized surfaces were rich in zinc, probably as ZnO or $Zn(OH)_2$, with 80/20 brass having less copper at the surface than 70/30 brass coupons.  The iron and chromium peaks shown in the polished 69.45% Cu spectra probably resulted from ion beam overlap onto the stainless steel carrousel.

Broad surveys were used primarily for qualitative analyses of surfaces. SIMS multiplex analyses were used to determine relative Cu/Zn levels at brass surfaces. In this technique, semi-quantitative measurements of surface copper content were made by cycling the SIMS analyzer over the mass range 60–70 amu while sputtering through a brass coating. SIMS data obtained from this narrow mass range were dumped onto magnetic tape after each cycle. Figure 9 shows a typical two-channel multiplex dump at a 200 $\mu$m spot on a brass coupon made by using PHI differentially pumped ion gun with no raster. These data were fed into an IBM 370/168 computer which corrected the $Cu_{63}$, $Cu_{65}$, $Zn_{64}$, $Zn_{66}$, $Zn_{67}$, $Zn_{68}$ peak intensities for background level, calculated the isotopic ratios $Cu_{63}/Cu_{65}$, $Zn_{64}/Zn_{66}$, $Zn_{64}/Zn_{68}$, $Zn_{66}/Zn_{68}$, compared the calculated isotopic ratios with theoretical ratios to check spectral interferences such as mass overlap, calculated $Cu_{63}/Zn_{64}$ ratio, and computed percent copper as a binary Cu + Zn alloy.

Quantitative surface analyses are commonly performed by applying elemental sensitivity factors to peak height data taken while sputtering. Hall[39] reported that an accuracy of about 30% could be expected from such an analytical treatment. Better accuracy is possible by using standards in the concentration range of the unknown which are analyzed under identical experimental conditions. Data from such standard calibration methods have been shown to fit a power series relationship as given by[40]

$$\% \; Cu \;\; = \;\; a \; R^b \qquad\qquad\qquad (5)$$

R  =  peak intensity of Cu at surface ($I^s_{Cu}$) divided by peak intensity of Zn at surface ($I^s_{Zn}$)

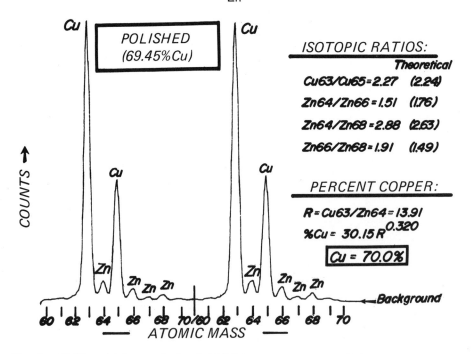

Fig. 9. SIMS multiplex analysis of 70/30 brass coupon at a 200 $\mu$m spot by using PHI differentially pumped ion gun with no raster.

Hall[39] showed that surface concentration ratio of a binary (A + B) alloy could be determined by applying a relative sensitivity factor ($S_r$) to peak intensities ($I^s$). According to

$$C_A^s / C_B^s = 1/S_r \ (I_A^s / I_B^s) \tag{6}$$

For a Cu + Zn alloy, Equation 6 becomes

$$\% \ Cu = 1/S_r R \tag{7}$$

Rearranging Equation 5 gives

$$\% \ Cu = a(R^{b-1}) \ R \tag{8}$$

or

$$1/S_r = aR^{b-1} \tag{9}$$

Therefore, sensitivity factors for copper and zinc vary with alloy compositions [39,40]. The constants a and b used in Equation 5 were calculated by measuring R from three well-characterized brass standards at 61.95% Cu, 72.29% Cu and 82.56% Cu. These standards were run each day because $R_{(\%Cu)}$ varied with elapsed time under vacuum[41]. By using Equation 5 and R = 13.91 as calculated in Fig. 9, SIMS multiplex analysis gave a copper level of 70.0% for a brass coupon having a bulk Cu content of 69.45% by X-ray fluorescence.

Quantitative analytical applications of SIMS must consider several different perturbation processes on a surface during ion bombardment, viz., preferential adsorption or desorption, generation and destruction of surface compounds, etc.[36,39]. For quantitative analysis of brass surfaces, e.g., steel tire cord, preferential sputtering, redeposition, or knock-on effects of one of the binary constituents (copper or zinc) are potential problems. By using theoretical isotopic distribution data, a $Cu_{63}/Zn_{64}$ ratio of 2.301 was calculated for a 61.95% Cu standard (see Table I), whereas actual SIMS measurements gave a ratio of 9.5. These results suggested that copper and zinc do not sputter in accordance with binary alloy composition ratio. Indeed, existence of sputtering irregularities as listed above were verified by using simultaneous SIMS + Auger multiplexing techniques[40]. Quantification of SIMS data by using relative sensitivity factors does not account for such irregularities. Table II lists copper contents for a 61.95% Cu standard calculated by using published relative sensitivity factors (MCL)[31] and factors calculated from the author research (GY)[42]. These values were too high, again illustrating the effects of sputtering on surface composition. Copper contents calculated from Equation 5 with constants derived from known standards analyzed under identical experimental conditions were, therefore, corrected for such sputtering anomalies. Also, combined surface analysis techniques were useful for detection of such potential problems.

Copper contents at the surface and after sputtering away approximately 500Å of brass are shown in Fig. 10 for as-polished and oxidized surfaces of both 70/30 and 80/20 coupons. These data were collected by using minimum size ion beams. Polishing appeared to leave a copper-rich surface and oxidization lead to zinc enrichment at the surface. Higher bulk copper 80/20 brass coupons showed more extreme changes in surface composition after both treatments. Figure 11 shows SIMS depth profiles obtained during combined SIMS/XPS/AES experiments where the ion beam was rastered over a 4mm x 4mm area in order to sputter etch a large enough surface area for XPS analyses. The data shown in these profiles were,

TABLE I

Theoretical ratios — copper & zinc

Given % Cu — 61.95

| Copper | Zinc |
|---|---|
| Mass 63  —  69.1% | Mass 64  —  48.9% |
| Mass 65  —  30.9% | Mass 66  —  27.8% |
|  | Mass 68  —  18.6% |

For Brass Standard

Mass 63 — 61.95% x 69.1%  =  42.81
Mass 64 — 38.05% x 48.9%  =  18.61
Mass 65 — 61.95% x 30.9%  =  19.14

R = 63/64 = 2.301
Cu = 63/65 = 2.236

TABLE II

Sensitivities

$$\frac{I_i X A_i}{S_i} + \frac{I_j X A_j}{S_j} + \ldots = 100$$

I  =  peak intensity
A  =  isotopic abundance
S  =  sensitivity factor

| I | A | $S_{GY}$ | $S_{MCL}$ |
|---|---|---|---|
| Cu: 91,424.05 | 69.1 | 42 | 32 |
| Zn: 16,395.05 | 48.9 | 12.9 | 20 |
| Standard=61.95%: | %Cu = | 70.8 | 83.1 |

therefore, averaged over a much larger area than given previously in Figs. 9 and 10, and, of course, the sputter rate was considerably decreased. However, these large area SIMS profiles agreed well with bulk copper level except for oxidized 80/20 brass which again showed zinc enrichment at the surface.

AUGER:  Broad-survey Auger analyses of oxidized 70/30 and 80/20 brass coupons using the larger beam electron gun (~100$\mu$m) of the PHI Model 548 system are shown in Fig. 12 for both as-received and sputter-cleaned surfaces. As-received surfaces showed traces of sulfur, chlorine, carbon, and nitrogen contamination in addition to copper, zinc, and oxygen. The 80/20 brass sample again showed much higher zinc enrichment at the surface as well as splitting of low-energy peaks indicating presence of oxide on the surface[43]. Sputtering removed the surface contaminants and most of the oxygen. Copper/zinc levels shown in the sputtered spectra were more representative of the bulk copper composition. Binary copper composition was corrected for sputtering artifacts by using a power curve relationship given by

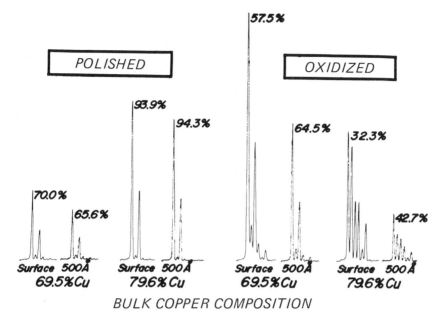

Fig. 10.  Copper content from 200μm spot SIMS multiplexing at surface and at 500Å for polished and oxidized 70/30 and 80/20 brass coupons.

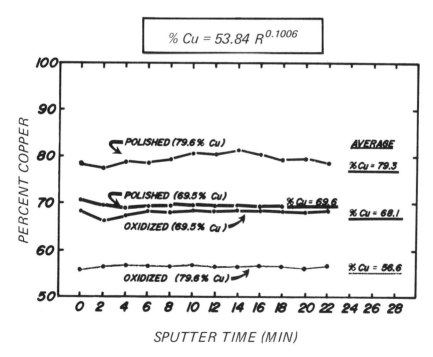

Fig. 11.  SIMS depth profiles for polished and oxidized 70/30 and 80/20 brass coupons calculated from SIMS multiplexing data obtained by rastering 3M ion gun over 4mm x 4mm area.

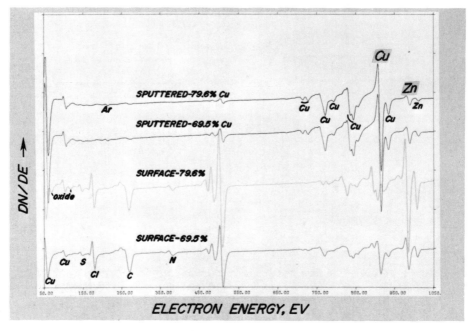

Fig. 12. Broad survey AES analyses using 100µm electron beam of oxidized 70/30 and 80/20 brass coupons at as-received surface and after sputtering.

$$\% \text{ Cu } = 42.74 \text{ R}^{0.250} \tag{10}$$

where R = peak-to-peak intensity ratio of Cu (920 eV) to Zn (994 eV). The Auger peaks used in Equation 10 are highlighted in Fig. 12.

Scanning Auger depth profiles of as-polished and oxidized 70/30 and 80/20 brass coupons are compared in Fig. 13. Plotted in Fig. 13 are atomic concentrations of Cu, Zn, O, S, N, and Cl calculated by MACS from Auger peak-to-peak intensities normalized by using relative sensitivity factors[44] versus argon sputter time. These Auger depth profiles were run simultaneously with SIMS profiles in Fig. 13 showed that surfaces of these coupons contained much lower copper than given by bulk composition. Remember that the atomic concentrations shown in these profiles were calculated from handbook relative sensitivity factors and so do not account for sputtering effects or changes in relative sensitivity with copper concentration. Nevertheless, several interesting features were seen in the SAM profiles. The polished 70/30 brass coupon showed a very small zinc oxide layer on the surface, whereas the polished 80/20 sample showed a small copper oxide layer. In fact, there was actually slight dezincification of the polished 80/20 brass coupon perhaps during the diamond polishing step. Both 70/30 and 80/20 oxidized samples showed zinc oxide at the surface; however, the 80/20 sample also showed a small copper oxide film on top of a thick zinc oxide film.

Surface chemistry calculated from both broad survey and profile Auger analyses for polished and oxidized 70/30 and 80/20 brass coupons is summarized in Table III. Initial surfaces of polished and oxidized 70/30 brass were not very different. Oxidized surfaces showed slightly higher oxygen content at the outermost surface, as well as traces of oxygen and carbon well into the bulk. On the other hand, polished and

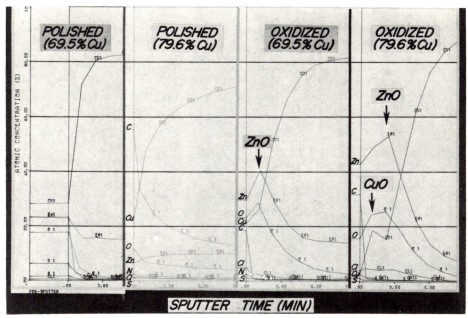

Fig. 13. SAM depth profiles of as-polished and oxidized 70/30 and 80/20 brass coupons.


TABLE III

Auger surface analyses: atomic concentration by using relative
sensitivities from handbook [40]

| | 70/30 Coupon (Cu+69.45%) | | | | 80/20 Coupon (Cu=79.60%) | | | |
|---|---|---|---|---|---|---|---|---|
| | Polished | | Oxidized | | Polished | | Oxidized | |
| Element | Surface (At %) | Bulk (At %) | Surface (At %) | Bulk (At %) | Surface (At %) | Bulk (At %) | Surface (At %) | Bulk (At %) |
| S | 3.5 | - | 0.7/0.6* | - | 0.5 | - | 0.5/0.7 | - |
| Cl | 0.3 | - | 3.0/4.2 | - | 0.7 | - | 2.3/2.8 | - |
| C | 26.6 | - | 22.0/20.6 | -/1.7 | 59.9 | 8.0 | 35.6/11.6 | -/1.9 |
| N | - | - | 1.4/2.5 | - | 1.3 | - | 17./1.3 | - |
| O | 10.3 | - | 13.7/19.5 | -/4.2 | 6.3 | 1.2 | 8.9/19.2 | -/1.8 |
| Cu | 32.8 | 83.9 | 27.1/23.4 | 84.8/79.9 | 23.8 | 81.1 | 2.8/9.0 | 91.1/86.4 |
| Zn | 26.4 | 16.1 | 32.0/29.2 | 15.2/14.3 | 7.4 | 9.7 | 48.2/55.2 | 8.9/9.8 |

*Underlined atomic percents from Auger survey analyses, all other values from Auger profile analyses.

R.M. SHEMENSKI

oxidized surfaces of 80/20 brass were very different. Polished coupons showed very high carbon on the surface extending well into the bulk. Copper level at surface of polished 80/20 brass was much higher than oxidized 80/20 brass, while zinc and oxygen were much lower on the surface of polished samples. Percent copper as a binary Cu + Zn alloy was calculated from the Auger data in Table III by using both relative sensitivities and peak ratios from known standards. These results, summarized in Table IV, show that quantitative analyses by the two techniques do in fact differ, in some cases by as much as 17%. These differences are even more apparent by plotting % $Cu_{(sen)}$ versus % $Cu_{(R)}$ in Fig. 14 where it is seen that % $Cu_{(sen)}$ was higher for Cu-rich alloys, lower for Zn-rich alloys, and approximately equal to %$Cu_{(R)}$ for a 50 Cu-50 Zn alloy. A comparison of % $Cu_{(sen)}$ and % $Cu_{(R)}$ for SIMS data is shown by a data bar in Fig. 14. The difference in these two values from SIMS data was quite similar to that found by Auger analyses. Therefore, the effect of sputtering on the analytical results was the same for both Auger and SIMS, which would not have been so had preferential sputtering occurred. The predominant element in the sputtered material shown by SIMS would have found to be deficient in the remaining surface as analyzed by Auger. This agrees with Wehner[15] who reported that differential sputtering is not a significant problem at ion beam energies greater than about 500 eV. However, Storp[45] found that zinc or perhaps ZnO was preferentially sputtered from the surface of 60/40 brass by using $Ne^+$, $Ar^+$ or $Xe^+$ at energies up to 5 keV. For 2.5–3.5 keV $Ar^+$ ions used in this study, redeposition and knock-in of the more concentrated element appear to adequately explain the results shown in Fig. 14. Whatever the physical nature of sputtering artifacts, error in quantification of surface analyses can be minimized by using Equation 5.

TABLE IV

Percent copper in binary Cu + Zn alloy from Auger analyses

| | 70/30 Coupon (Cu=69.45%) | | | | 80/20 Coupon (Cu=79.60%) | | | |
|---|---|---|---|---|---|---|---|---|
| | Polished | | Oxidized | | Polished | | Oxidized | |
| | Surface (At %) | Bulk (At %) | Surface (At %) | Bulk (At %) | Surface (At %) | Bulk (At %) | Surface (At %) | Bulk (At %) |
| $R = \dfrac{I_{Cu(920)}}{I_{Zn(994)}}$ | 1.61 | 6.76 | 1.10/1.04* | 7.20/7.22 | 4.17 | 10.9 | .07/0.2 | 13.25/11.38 |
| $\%\,Cu = aR^b$ | 48.1 | 68.9 | 43.7/43.1 | 70.0/70.1 | 61.1 | 77.6 | 22.3/29.0 | 81.6/78.5 |
| $\%\,Cu = \dfrac{I_{Cu}/S_{Cu}}{I_{Cu}/S_{Cu} + I_{Zn}/S_{Zn}}$ | 55.4 | 83.9 | 45.8/44.5 | 84.8/84.8 | 76.3 | 89.4 | 5.4/14.1 | 91.1/89.8 |

*Underlined values from Auger survey analyses, all other values from Auger profile analyses

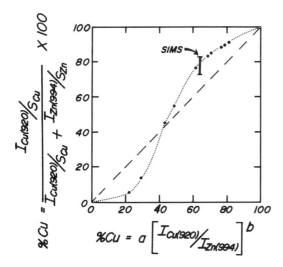

*(% Cu from handbook sensitivities versus measured ratios)*

Fig. 14. Comparison of quantification of Auger data by using published sensitivities versus peak height ratios measured from known standards.

XPS: Broad survey XPS analyses of as-prepared and sputter-cleaned 70/30 and 80/20 brass surfaces are shown in Fig. 15. The locations of 8 peaks for both zinc and copper (4 photoelectron and 4 Auger electron peaks), an oxygen peak, and a carbon peak are shown in XPS spectra used for analysis of brass surfaces. Just as was found by SIMS and Auger analyses, oxidized surfaces showed higher zinc, carbon, and oxygen compared with a clean, sputtered surface. However, these differences along with peak shifts due to different chemical combinations or valence states are shown more dramatically in narrow-range surveys, Figs. 16–21. Figures 16 and 17 show XPS spectra from 250–350 eV for 70/30 and 80/20 brass, respectively. This range of binding energy includes the principal copper Auger peak ($CuL_3VV$), the principal zinc Auger peak ($AnL_3M_{45}M_{45}$), and the carbon 1s photoelectron peak. XPS analyses of oxidized 70/30 brass (Cu = 69.45%) showed three zinc Auger peaks which remained up to 50 minutes sputtering. These peaks were identified as $Zn(OH)_2$, ZnS, and $Zn^o$[46]. By using an Auger electron in the XPS spectra, copper was not detected until 55 minutes sputtering, at which time copper Auger peaks identified as CuO and $Cu^o$[46] were seen. Figure 17 shows similar results after sputtering an oxidized surface of 80/20 brass. However, for oxidized 80/20 brass, a ZnO peak was also identified in these spectra up to 110 minutes sputtering. This ZnO appeared to be the major zinc constituent at the as-prepared surface of oxidized 80/20 brass, whereas ZnS was the largest peak for 70/30 brass. Copper as CuO and $Cu^o$ appeared at 110 minutes sputtering for 80/20 brass. After 210 minutes sputtering, the $Cu^o$ peak had increased relative to CuO and the zinc Auger peak had almost totally disappeared.

Figures 18 and 19 show narrow-range survey spectra (390–540 eV) for 70/30 and 80/20 brass surfaces. This range included minor Auger peaks of copper and zinc ($CuL_3M_{23}M_{23}$, $ZnL_3M_{23}M_{23}$, $CuL_3M_{23}V$) and the oxygen 1s photoelectron peak.

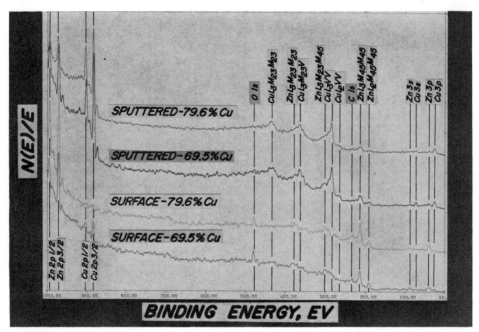

Fig. 15. XPS broad survey spectra from oxidized and sputtered-clean surfaces of 70/30 and 80/20 brass coupons.

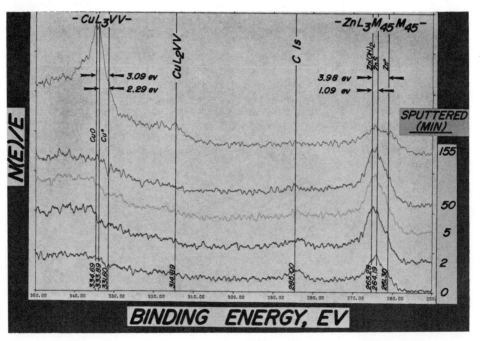

Fig. 16. XPS narrow-range survey spectra (250–350 eV) of oxidized 70/30 brass sputtered for 0, 2, 5, 50, and 155 minutes.

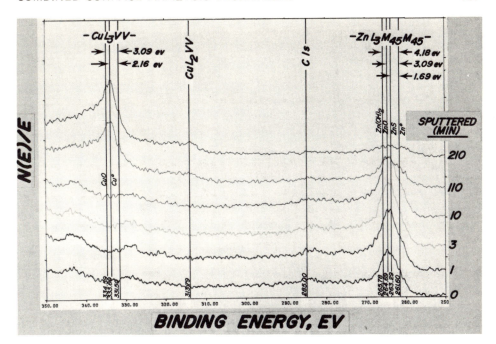

Fig. 17. XPS narrow-range survey spectra (250–350 eV) of oxidized 80/20 brass sputtered for 0, 1, 3, 10, 110, and 210 minutes.

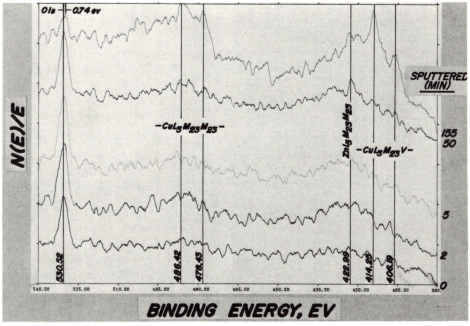

Fig. 18. XPS narrow-range survey spectra (390–540 eV) of oxidized 70/30 brass sputtered for 0, 2, 5, 50 and 155 minutes.

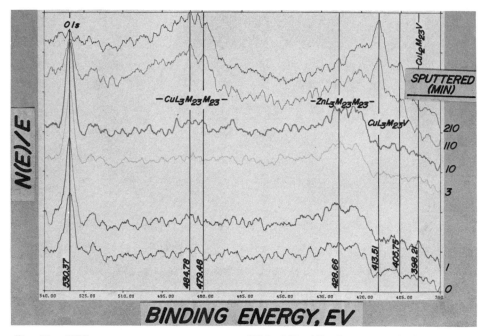

Fig. 19.  XPS narrow-range survey spectra (390–540 eV) of oxidized 80/20 brass sputtered for 0, 1, 3, 10, 110, and 210 minutes.

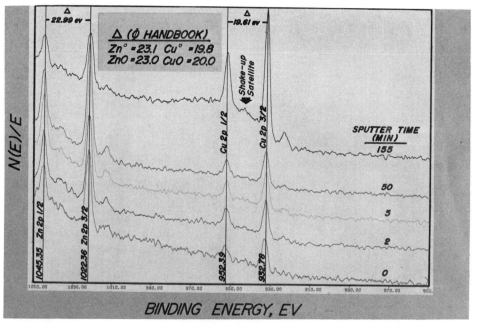

Fig. 20.  XPS narrow-range survey spectra (850–1050 eV) of oxidized 70/30 brass sputtered for 0, 2, 5, 50, and 155 minutes.

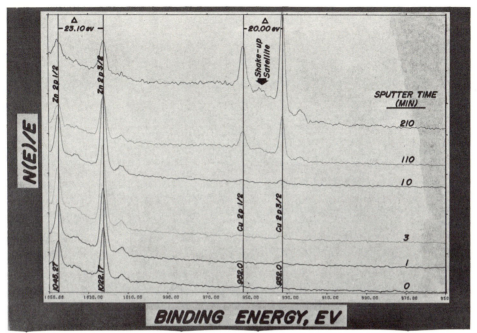

Fig. 21. XPS narrow-range survey spectra (850–1050 eV) of oxidized 80/20 brass sputtered for 0, 1, 3, 10, 110, and 210 minutes.

The copper and zinc Auger peaks varied with sputter depth identically to the lower binding energy Auger peaks shown in Figs. 16 and 17. Location of the oxygen 1s peaks was the most important information obtained from Figs. 18 and 19. The oxygen 1s peak for 80/20 brass was shifted to a lower binding energy than for 70/30 brass. A shift of oxygen 1s peak to lower binding energy with increased oxidation of copper was reported by Schon[47], and so XPS also showed that 80/20 brass was more oxidized.

Narrow-range XPS surveys (850–1050 eV) shown in Figs. 20 and 21 include prominent 2p3/2 and 2p1/2 photoelectrons for copper and zinc. Note that oxidation caused only a very slight chemical shift of the 2p peaks[45,48] so that these peaks could not be used to identify various chemical species present on the surface as could the Auger peaks in Figs. 16 and 17. However, oxides are indicated by presence of shake-up satellites in the copper 2p spectrum[49]. Small shake-up satellites are shown in Figs. 20 and 21. Due to lower kinetic energy of 2p electrons, their escape depth (mean escape depth about 7Å in metallic Cu/Zn[45]) differs considerably from that of Auger electrons (about 10Å ), or 3d electrons (approximately 16Å) [50,51]. Therefore, the surface depth analyzed by XPS is about 20 Å in the low energy range (2p electrons) and about 50Å in the high energy range (3d electrons). This sampling depth may be increased by a factor of 2 for oxides[45].

Both copper and zinc were found at the as-prepared surface and after 155 minutes sputtering of oxidized 70/30 brass (see Fig. 20). Conversely, Fig. 21 showed no copper at the as-prepared surface and much less zinc after sputtering oxidized 80/20 brass. These data agreed very well with SIMS data shown in Fig. 10 and Auger data given in Fig. 12. Comparison on the separation of 2p3/2–2p1/2 photoelectron peaks for copper and zinc shown in Figs. 20 and 21 with PHI handbook data[49] showed the as-prepared oxidized surface of 70/30 brass to contain zinc oxide plus metallic copper and 80/20 brass to contain copper oxide plus metallic zinc. These results again agreed directly with Auger profile data shown in Fig. 13. XPS binding energies for photoelectron and Auger electron peaks found in this study are summarized in Table V for both 70/30 and 80/20 brass samples.

TABLE V

XPS binding energies of photoelectron and auger peaks
for oxidized 70/30 and 80/20 brass

| XPS Peak | Binding Energy (eV) | |
|---|---|---|
| | 70/30 Brass - Oxidized | 80/20 Brass - Oxidized |
| CuL$_3$VV | | |
| CuO | 334.7 | 334.6 |
| Cu$^o$ | 333.9 | 333.7 |
| ZnL$_3$M$_{45}$M$_{45}$ | | |
| Zn(OH)$_2$ | 265.3 | 265.8 |
| ZnO | - | 264.7 |
| ZnS | 264.2 | 263.3 |
| Zn$^o$ | 261.3 | 261.6 |
| Cu 2p 3/2 | 932.8 | 932.0 |
| Cu 2p 1/2 | 952.4 | 952.0 |
| △ | 19.6 | 20.0 |
| Zn 2p 3/2 | 1022.4 | 1022.2 |
| Zn 2p 1/2 | 1045.4 | 1045.3 |
| △ | 23.0 | 23.1 |
| O 1s | 530.5 | 530.4 |

Quantification of XPS data was accomplished by analyzing known standards to derive calibration equations as given by Equation 5 for three photoelectrons, viz., 2p3/2, 2p1/2 and 3p, and LMM Auger electrons. These XPS copper analyses of oxidized 70/30 and 80/20 brass are compared with SIMS and Auger results in Fig. 22. For 60 Cu/40Zn brass alloys, Storp[45] reported that quantity ratio (Cu/Zn) was 1.1–1.2 times the intensity ratio (R) for XPS 2p photoelectrons. Rearranging this relationship gives

$$\% \ Cu = \frac{1.15 \ R}{1 + 1.15R} \ X \ 100 \qquad (11)$$

Copper profiles in polished and oxidized 70/30 and 80/20 brass coupons were calculated by using Equation 11 with XPS peak intensities for 2p, 3p and LMM electrons as shown in Fig. 23. These values agreed reasonably with copper levels calculated from Equation 5 (see Fig. 22). More importantly, however, these copper profiles determined from 2p, 3p, or LMM electrons were very similar; the slight variations observed in Fig. 23 could be attributable to different sensitivity factors and experimental error in peak measurement. Escape depths of the XPS electrons used decrease in the order 3p (~16 Å), LMM (~10Å), and 2p (~7 Å), and, therefore, XPS by using 2p photoelectrons would be more surface sensitive than for 3p photoelectrons. If preferential sputtering of zinc or redeposition, knock-in, etc., of copper occurred, an apparent accumulation of copper would be seen in the region of the very

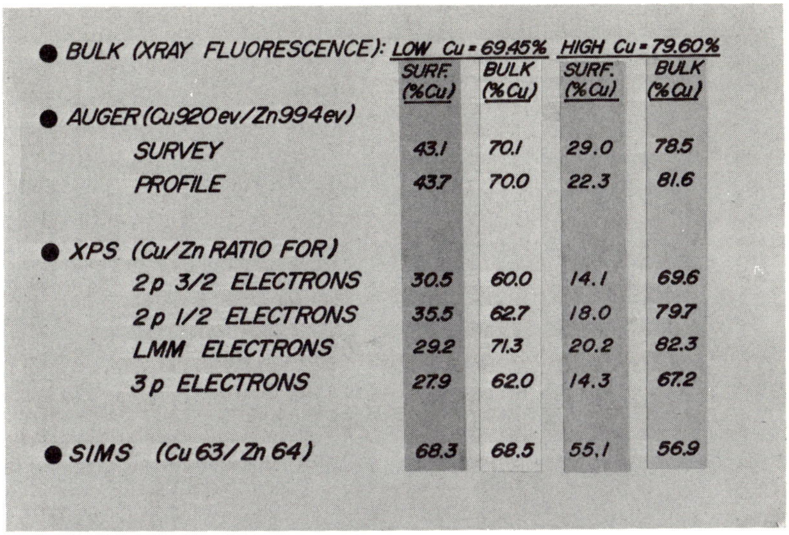

| ● BULK (XRAY FLUORESCENCE): | LOW Cu = 69.45% | | HIGH Cu = 79.60% | |
|---|---|---|---|---|
| | SURF. (%Cu) | BULK (%Cu) | SURF. (%Cu) | BULK (%Cu) |
| ● AUGER (Cu 920 ev / Zn 994 ev) | | | | |
| SURVEY | 43.1 | 70.1 | 29.0 | 78.5 |
| PROFILE | 43.7 | 70.0 | 22.3 | 81.6 |
| | | | | |
| ● XPS (Cu/Zn RATIO FOR) | | | | |
| 2p 3/2 ELECTRONS | 30.5 | 60.0 | 14.1 | 69.6 |
| 2p 1/2 ELECTRONS | 35.5 | 62.7 | 18.0 | 79.7 |
| LMM ELECTRONS | 29.2 | 71.3 | 20.2 | 82.3 |
| 3p ELECTRONS | 27.9 | 62.0 | 14.3 | 67.2 |
| | | | | |
| ● SIMS (Cu 63/ Zn 64) | 68.3 | 68.5 | 55.1 | 56.9 |

Fig. 22. SIMS/AES/XPS surface analyses of oxidized brass coupons calculated from copper/zinc peak height ratios.

thin emission zone of the neutral particles and secondary ions[15], and would contribute proportionately more the the 2p XPS signal than to the 3p signal. Such effects were not found in Fig. 23.

Another test for occurrence of sputtering artifacts is to plot ratio of 2p/3p photoelectrons for both copper and zinc as a function of sputtering time. An increase in this 2p/3p ratio indicates surface enrichment, i.e., redeposition or knock-in; whereas, preferential sputtering would be indicated by a decreased 2p/3p ratio. Figure 24 shows 2p 3/2 /3p photoelectron ratio for copper and zinc from polished and oxidized 70/30 and 80/20 brass as a function of sputter time. A simultaneous increase in the 2p/3p ratio for both copper and zinc near the surface indicated removal of surface contamination or oxide films. The 70/30 brass coupons showed no evidence of surface perturbations due to sputtering, except for perhaps slight copper redeposition after long sputtering times of oxidized 70/30 brass. The 2p/3p ratios from higher copper 80/20 brass alloys indicated a somewhat greater tendency for sputtering anomalies, probably again copper redeposition as shown by a decreasing 2p/3p ratio for zinc with a corresponding increase in the copper ratio.

SUMMARY

Determination of the quality of brass surfaces by using specialized spectroscopic analytical techniques, viz., secondary ion mass spectrometry (SIMS), Auger electron spectroscopy (AES), and X-ray photoelectron spectroscopy (XPS) was demonstrated.

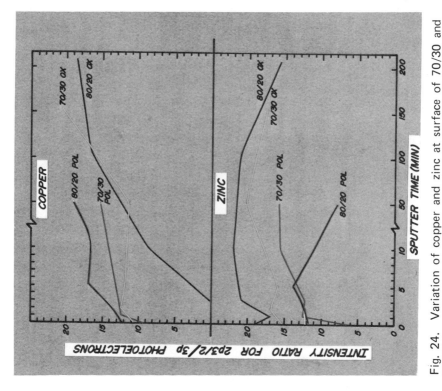

Fig. 24. Variation of copper and zinc at surface of 70/30 and 80/20 brass upon sputtering with 2.5 keV Ar+ as determined by ratio of 2p3/2 / 3p XPS peaks.

Fig. 23. Cu/Zn intensity ratios calculated from 2p3/2, 3p, and LMM XPS peaks as a function of sputter time for 70/30 and 80/20 brass coupons.

## TABLE VI

Copper concentrations in brass alloys by surface analyses techniques

| Sample | SIMS | | AES/SAM | XPS | | | |
|---|---|---|---|---|---|---|---|
| | | | | | 2p3/2 | 2p1/2 | 3p | LMM |
| 70/30 Polished | Surface: | 70.0% | Surface: | | | | |
| | 500Å: | 65.6% |   Ratio - 48.1% | Surface: | 72% | 68% | 60% | 80% |
| | Profile (Avg): | 69.6% |   Sensit- 55.4% | | | | | |
| | | | Bulk: | Bulk: | 76% | 76% | 77% | 86% |
| | | |   Ratio - 69.9% | | | | | |
| | | |   Sensit- 83.9% | | | | | |
| 70/30 Oxidized | Surface: | 93.9% | Surface: | Surface: | 30% | 29% | 35% | 39% |
| | 500Å: | 94.3% |   Ratio - 43.7%/43.1% |   Ratio - | 30.5% | 35.5% | 27.9% | 29.2% |
| | Profile (Avg): | 68.1% |   Sensit- 45.8%/44.5% | | | | | |
| | | | Bulk: | Bulk: | 70% | 64% | 80% | 66% |
| | | |   Ratio - 70.0%/70.1% |   Ratio - | 60.0% | 62.7% | 62.0% | 71.3% |
| | | |   Sensit- 84.8%/84.8% | | | | | |
| 80/20 Polished | Surface: | 57.5% | Surface: | Surface: | 68% | 60% | 76% | 67% |
| | 500Å: | 64.5% |   Ratio - 61.1% | | | | | |
| | Profile (Avg): | 79.3% |   Sensit- 76.3% | Bulk: | 88% | 84% | 91% | 74% |
| | | | Bulk: | | | | | |
| | | |   Ratio - 77.6% | | | | | |
| | | |   Sensit- 89.4% | | | | | |
| 80/20 Oxidized | Surface: | 32.3% | Surface: | Surface: | 0 | 0 | 0 | 0 |
| | 500Å: | 42.7% |   Ratio - 22.3%/29.0% |   Ratio - | 14.1% | 18.0% | 14.3% | 20.2% |
| | Profile (Avg): | 56.6% |   Sensit- 5.4%/14.1% | | | | | |
| | | | Bulk: | Bulk: | 75% | 73% | 88% | 72% |
| | | |   Ratio - 81.6%/78.5% |   Ratio - | 69.6% | 79.7% | 67.2% | 82.3% |
| | | |   Sensit- 91.1%/89.8% | | | | | |

## TABLE VII

Summary of relative advantages and disadvantages
of SIMS/AES/XPS surface analytical techniques

| Technique | Advantages | Disadvantages |
|---|---|---|
| SIMS | $10^{-6}$ to $10^{-9}$ detection limit<br>Isotopic & ion fragment information<br>Fast<br>Monolayer depth resolution<br>All elements from H to U<br>Spatial resolution none to $1-2\mu m$<br>  depending on beam diameter | Not quantitative<br>Large matrix effects<br>Sputtering anomalies<br>Limited chemical information |
| AES | Minimal matrix effects<br>Fast<br>Spatial resolution $5-10\mu m$<br>Sensitive to low Z elements<br>10-30Å depth resolution<br>All elements from Li-U | Detection limit $\sim 0.1\%$<br>Electron beam degradation<br>Insulator problems<br>Large spectral background<br>Chemical information limited |
| XPS | Chemical & valence state information<br>Low background<br>No insulator problems<br>Negligible X-ray beam effects<br>10-30Å depth resolution<br>All elements from Li-U | Large area analyses<br>Very poor spatial resolution<br>Slow<br>Detection limit $\sim 0.1\%$ |

SIMS was useful for qualitative information of elements present on a surface. SIMS detected elements over a very large mass range (hydrogen-uranium) and at very low trace amounts ($10^{-6}$-$10^{-9}$ atomic fraction). Also, limited chemical information was obtained from ionic fragments, combinations, and isotopes found in SIMS spectra. By using well-characterized known standards, semi-quantitative SIMS depth profile analyses were made through brass surfaces. AES/SAM techniques gave quantitative information accurate to about 20–30% of elements present in brass surfaces at levels greater than one atomic percent[52]. These techniques were useful for spatial resolution, e.g., segregation, conglomeration, partitioning, etc., of elements on a brass surface and provided some very limited chemical information based on Auger peak shapes. XPS gave only average surface chemistry information over a large area (approximately 3 mm in diameter). However, this technique provided unique information on chemical combinations or valence states present on brass surfaces, such as zinc present as $Zn(OH)_2$, $ZnO$, $ZnS$, or $ZnO$. Due to different escape depths of the high (3p) and low energy (2p) copper and zinc photoelectrons in an XPS spectrum, differences in chemical composition of the outermost atomic layers could be distinguished from "bulk" chemistry. Therefore, for processes or products where surface quality is a critical factor in overall product reliability, such as brass-coated steel tire cord, surface analysis techniques are ideally suited for quality control.

All three of these surface analysis techniques provided semi-quantitative copper concentrations of brass surfaces and depth profiles by using a power-series relationship given in Equation 5. These data are summarized in Table VI along with % Cu calculated from handbook sensitivities (AES) and Equation 11 (XPS). Although there was very good general agreement among the chemical analyses given in Table VI for different surface analysis techniques, there were also large discrepancies. For example, SIMS gave 93.9% Cu at the surface of oxidized 70/30 brass compared with 30–40% Cu for AES and XPS. Also, a "bulk" analysis for oxidized 80/20 brass of 56.6% Cu was calculated from SIMS data, whereas AES gave 80–90%, and XPS gave 70–90%. It is clear from these data and from the discussion above that each technique has distinct capabilities and limitations, strengths and weaknesses. A comparison of the relative advantages and disadvantages for these three surface analysis techniques is summarized in Table VII[18,26,52].

No single technique is a panacea for all surface analysis problems. Certainly, the limitations listed in Table VII dictate using a combination of two or more of these techniques, preferable simultaneous measurements as used in this study. Combination techniques, especially by using simultaneous, *in-situ* measurements are more effective in that direct correlation of SIMS/AES/SAM/XPS spectra from the same sample area permits confirmation of experimental data by two or more independent methods[53]. Also, combining SIMS with AES, SAM, or XPS permits a better understanding of the sputtering process since both the sputtered secondary ions as well as the remaining surface are concurrently analyzed[54]. Combination techniques also provide an economic advantage since a single ultra-high vacuum chamber and data storage and processing system are used, and some of the electronic components can be shared. In addition, since data are accumulated concurrently, savings are realized in analytical time.

CONCLUSION

Surface analysis presents a new dimension to quality control and opens up areas previously inaccessible to comprehensive analytical regulation. Quality control of surfaces is most reliably provided by combined techniques which reinforce the advantages of each individual method so as to overcome their disadvantages.

REFERENCES

1. "ASTM Manual on Quality Control of Materials," STP No 15–C, Jan (1951).
2. "Source Book in Failure Analysis," Amer. Soc. for Metals, Oct. (1974).
3. T.J. Dolan, *Metals Eng Quart, 12,* p. 32, Nov. (1972).
4. J.S. Brinen, *Acc of Chem Res, 9,* p. 86 (1976).
5. J.M. Morabito & R.K. Lewis in "Methods of Surface Analysis," ed. A.W. Czanderna, Elsevier Scientific, New York (1975).
6. E. Debruyne & A. Maesseele, *Rubber Chem & Tech, 42–2,* p. 613, (1969).
7. R. Halsberghe, *Wire Jour, 10,* p. 65 (1976).
8. O.A. Drica-Minieris, *Wire Jour, 11,* p. 60 (1978).
9. P.F. Murray, *Wire Jour, 4,* p. 50 (1977).
10. W.J. van Ooij, *Wire Jour, 8,* p. 40 (1978).
11. G. Haemers, *Wire Jour, 9,* p. 134 (1978).
12. W.J. van Ooij, *Surf Tech, 6,* p. 1 (1977).
13. W.J. van Ooij, *Surf Sci, 68,* p. 1 (1977).
14. W.J. van Ooij, *Rubber Chem & Tech, 3/4,* p. 52 (1978).
15. G.K. Wehner in "Methods of Surface Analysis," ed. A.W. Czanderna, Elsevier Scientific, New York, p. 5 (1975).
16. D. Lichtman, *ibid.,* p. 39.
17. W.J. van Ooij, to be published in 1979 *Rubber Reviews of Rubber Chem & Tech.*
18. C.A. Evans, Jr., *Anal Chem, 47,* p. 818 (1975).
19. J.R. Cuthill, *ASTM Standardization News, 6,* p. 8 (1975).
20. R.S. Carbonara, *ibid,* p. 12.
21. C.J. Powell, *ibid,* p. 15.
22. N.E. Erickson, *ibid,* p. 18.
23. J.T. McKinney, *ibid,* p. 21.
24. B.F. Phillips, *ibid,* p. 25.
25. G.C. Nelson, *ibid,* p. 28.
26. N.A. Alford, A. Barrie, I.W. Drummond, & Q.C. Herd, *Surf & Interface Anal, 1,* p. 36 (1979).
27. R.S. Swingle & W.M. Riggs, *Crit Rev Anal Chem, 5,* p. 267 (1975).
28. R.E. Weber, *Res/Dev, 23,* p. 22 (1972).
29. F.W. Karasek, *Res/Dev, 24,* p. 25 (1973).
30. G.R. Sparrow, *Res/Dev, 27,* (1976).
31. G.R. Sparrow, presented to Amer Soc for Mass Spect, San Diego, Calif, May (1976).
32. J.A. McHugh, in "Secondary Ion Mass Spectrometry", NBS Spec Pub #427, p. 129 (1974).
33. J.M. Schroer, *ibid,* p. 121.
34. H.W. Werner, in "Electron & Ion Spec of Solids," eds. L. Fiermans, J. Vennek, & W. Oekeyser, Plenum Press, p. 324 (1978).
35. G. Blaise, in "Material Characterization Using Ion Beams," eds. J.P. Thomas & A. Cachard, Plenum Press, p. 164 (1978).
36. A. Benninghoven, *Surf Sci, 53,* p. 596 (1975).
37. B.N. Colby and C.A. Evans, Jr, *App Spect, 27,* p. 274 (1973).
38. T. Ishitani and H. Tamura, *Surf Sci, 55,* p. 179 (1976).
39. P.M. Hall & J.M. Morabito, *Surf Sci, 62,* p. 1 (1977).
40. R.M. Shemenski, S.J. Vonk, & E.F. Riggenbach, presented to Amer Chem Soc – Rubber Div, Oct. (1977).
41. R.M. Shemenski, presented to 3M Users' Conf, Sept. (1978).
42. R.M. Shemenski, unpublished results.
43. G.T. Burstein & T.P. Hoar, *Corr Sci, 18,* p. 75 (1978).
44. Physical Electronics Industries, Inc, "Handbook of Auger Electron Spectroscopy" (1976).

45. S. Storp and R. Holm, *Proc 7th Internat Vac Congr on Solid Surf,* Vienna (1977).
46. D. Benko, unpublished research.
47. G. Schon, *Surf Sci, 35,* p. 96 (1973).
48. R. Holm and S. Storp, *Appl. Phys, 9,* p. 217 (1976).
49. Physical Electronics Industries, Inc, "Handbook of X—Ray Photoelectron Spectroscopy" (1979).
50. C.R. Brundle, *Jour Vac Sci Tech, 11,* p. 212 (1974).
51. D.R. Penn, *Jour Elect Spect, 9,* p. 29 (1976).
52. J.M. Morabito, in "Secondary Ion Mass Spectrometry," NBS Spec Pub #427, p. 191 (1974).
53. S. Komiya and T. Narusawa, *Proc 7th Vac Congr on Solid Surf,* Vienna (1977).
54. G. Kaus and J. Kempf, *ibid.*

# METALLOGRAPHIC CONTROL IN POWDER METALLURGY

G. Elssner*, W. Kaysser*, and G. Petzow*

## INTRODUCTION

One of the stringent requirements for the economical production of sintered metal and ceramic parts is a metallographic control of their different manufacturing stages. In view of the development of new and improved powder metal products metallographic investigation is a powerful tool for determining important details of sintering mechanisms. In both areas advanced and conventional metallographic methods exist which have to be fitted to the special requirements of powder metallurgy. Stereology combined with metallography serves to establish quantitative correlations between microstructure and mechanical and physical properties of the starting powder, the compact and the sintered part. Qualitative metallography based on incident light microscopy and scanning electron microscopy is as yet the most commonly used method to control microstructural features in powder metallurgy. Progress in preparation and contrasting techniques offers the chance of a rather quick displacement of conventional qualitative methods by quantitative microstructural analysis of sintered materials.

This paper mainly considers some of the metallographic methods employed in investigations on liquid phase sintering to characterize the microstructure and its influence on materials behavior and properties. The selected examples aim to demonstrate the large extent of applicability of metallography to powder metallurgy problems in research and development as well as in production control.

## CONTROL OF POWDERS AND COMPACTS

Particle shape and the distribution of particle sizes of a powder determine predominantly its behavior during processing into compacts. The transmitted light microscopy technique is the common method to characterize metal powder by the shape and the size of its dispersed particles which, however, are only visible by their outlines. More information is attainable by SEM inspection of powder particles. The excellent depth resolution of this type of microscope reveals details on shape and surface structure both of the finest particles and of coarse agglomerates. A complete qualitative powder characterization also includes the examination of polished sections of resin embedded powder to get information of the internal microstructure of particles, aggregates and agglomerates. Figure 1 gives a simple example of a quantitative description of the size distribution of iron spheres which were used for model liquid phase sintering experiments[1]. Semi-automatic image analysis methods are nowadays available allowing a simultaneous qualitative evaluation of size and shape distribution of powders[2].

*Max-Planck-Institut fur Metallforschung, Institut fur Werkstoffwissenschaften, Stuttgart, W. Germany.

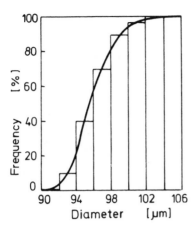
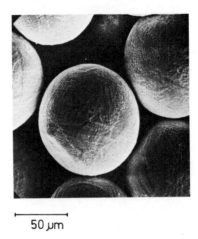

Fig. 1.   Diameter distribution and SEM micrograph of spherical iron powder.

Metallographic inspection of polished sections from compacts gives insight into their pore size and shape distribution and yields information on the degree of homogeneity of mixed powder compacts after pressing and before sintering.   To diminish preparation difficulties arising from their high porosity, on the order of 10 to 40 percent, it is advisable to presinter the compacts slightly before cutting.   Figure 2 shows the intercept distribution of pores in a W–Cu compact before and after liquid phase sintering.   In this model experiment with metal spheres ellipsoidal pores were found after pouring.   Therefore, the distribution of intercept length measured parallel and perpendicular to the axis were different.   After sintering the pores became spherical and the intercept distributions are equal for all directions.   This type of measurement allowed a semiquantitative explanation for macroscopic shrinkage anisotropy during sintering[3].

## CONTROL OF THE SINTERING PROCESS

Modern metallography is not restricted to the analysis or description of static microstructures.   For powder metallurgical purposes hot-stage scanning electron microscopy can be employed to observe *in situ* dynamic microstructural changes during sintering[4,5].   Loose three dimensional packings of mixtures of both W–Cu and Fe–Sn spheres having 10 w/o Cu or Sn were used to investigate the rearrangement process and subsequent stages of liquid phase sintering.   The changes in microstructure were recorded on video-tape (30 half frames per second) or by a series of single micrographs.   Figures 3a to d show a region with an initially high melt content of a Fe–Sn specimen during an annealing treatment at $450°C$.   After 3 min. the melt has penetrated into the Fe particles.   The onset of the formation of $FeSn_2$ crystals on the surface of the iron particles is visible after 15 min.   The growth of these crystals and the consumption of the melt can be seen in the micrographs corresponding to annealing times of 50 and 90 min.   Hot-stage SEM micrographs furnish information on the development of surface microstructural features.   Examination of microstructures obtained by metallographic sectioning from the interior of the specimens were used to confirm the results observed during hot-stage SEM analysis[5].   In addition Fig. 4 shows the microstructure of a Fe–10 w/o Sn compact

Anisotropic particle rearrangement

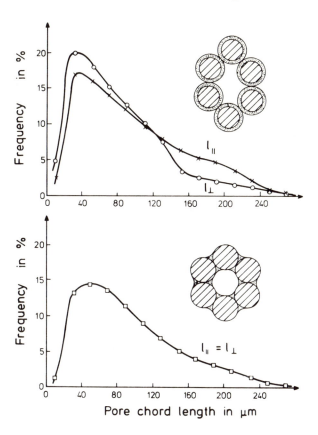

Fig. 2.  Intercept distribution of pores in W–Cu specimens before and after sintering[3].

pressed at 785 MPa and annealed for 1 min. at 450°C.  Due to a tighter packing of spheres only a small amount of the melt was able to penetrate into the areas between the particles.  Generally 10 percent of the pressed contacts were filled totally and 10 percent filled partially with melt.  Figure 4 is from a region of pressed spheres filled with a large amount of melt.  The micrograph was taken from a polished section which was additionally gaseous ion coated[6].  For powder metallurgy the techniques of interference layer microscopy proved to be especially suited to the differentiation of phases for both heterogeneous and porous microstructures.

Special etching techniques[7] were used to study shape accommodation and coalescence of tungsten grains during liquid phase sintering[8].  In a model experiment spherical tungsten particles (single crystals, 200 to 250 $\mu$m in diameter) were mixed with fine (10$\mu$m) tungsten powder and 4 w/o of fine nickel powder (less than 44$\mu$m), and sintered at 1670°C in a $H_2$ atmosphere.  Polished sections of the sintered mixture were etched in Murakami's solution for a few minutes.  Figure 5 shows the

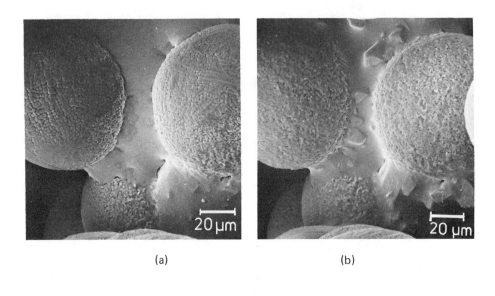

(a)                                      (b)

(c)                                      (d)

Fig. 3. Growth of the FeSn$_2$ phase in a loose packing of Fe–10 w/o Sn at 450°C observed in a hot-stage SEM.  a. Annealing time 3 min.  b. Annealing time 15 min. c. Annealing time 50 min.  d. Annealing time 90 min.

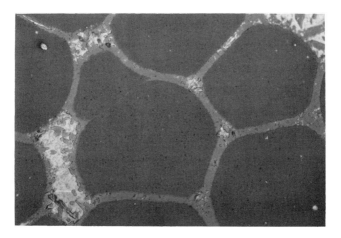

Fig. 4.  Microstructure of a Fe—10 w/o Sn specimen pressed at 785 MPa and annealed at 450°C for 1 min.  600x.  (Reduced 35% for reproduction.)

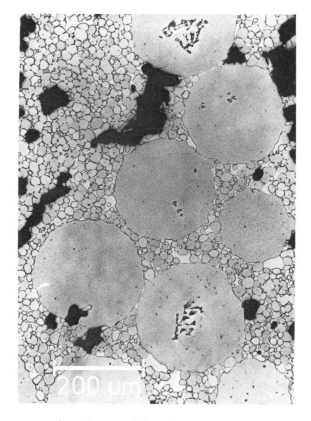

Fig. 5.  Microstructure of a mixture of 48 w/o large tungsten spheres, 48 w/o tungsten powder and 4 w/o nickel sintered for 3 min. at 1670°C[8].

microstructure after 3 min. of sintering at 1670°C. Both large spherical and rounded fine tungsten particles are visible. The porosity is 17%. After 2 hr. of sintering a completely different microstructure was generated with only 2% porosity (Fig. 6). An "overetching" method was chosen to reveal spherical cores of the pure original tu tungsten particles surrounded by tungsten-rich solid solution layers (0.15 w/o Ni in W) forming the typical contact flattened heavy alloy structure. A comparison of Fig. 6 with Fig. 5 strongly suggests that the small original tungsten particles dissolve into the liquid and reprecipitate on the surface of the large tungsten particles. Theoretical considerations[9] based on metallographic investigations lead to the conclusion that the main material flux during liquid phase sintering is determined by diffusion controlled Ostwald ripening. Figure 7 presents a model for a mixture of large and small particles together with the well known Thomson-Freundlich equation ($\mu-\mu_0$ = chemical potential difference, c = solubility of W in the melt at the surface of a W particle of radius r, $c_0$ = equilibrium solubility of tungsten in the melt, $\gamma_{SL}$ = solid liquid interfacial energy, $\Omega$ = molar volume). Grain shape accommodation and densification is favored by grain growth due to Ostwald ripening and by the action of densification forces due to the presence of pores.

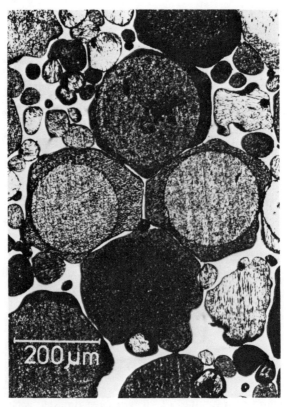

Fig. 6.   Microstructure of the W–4 w/o nickel mixture sintered for 2 hr. at 1670°C[8].

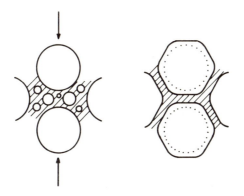

Rearrangement controlled by Ostwald
ripening , $\mu - \mu_0 = RT \ln \dfrac{c}{c_0} = \dfrac{2 \gamma_{SL}}{r} \Omega$

Fig. 7. A model for contact flattening during liquid phase sintering[8].

## CONTROL OF SINTERED PARTS

The most significant feature of sintered ceramic and metal parts is the degree
of porosity. The mechanical properties and other important properties of sintered
products depend to a great extent on the size, shape, distribution, and amount of
total porosity. At present, the difficulties of automatic or semi-automatic image
analysis of porous microstructures arise mainly from the imperfections in the art of
specimen preparation. Therefore, for embedded sections of sintered ceramic materials preparation methods were worked out using automatic equipment and diamond
as a grinding and polishing agent[10,11]. The preparation time for high quality
polished surfaces could be reduced for $Al_2O_3$, $ZrO_2$ SiC, and other ceramics to 24
to 40 min (Fig. 8). The reproducibility of the quality of the sections was checked
by porosity measurements. As an example the surface of an $Al_2O_3$ ceramic obtained from cutting and fissured with grain pop outs is shown in Fig. 9. After a
total preparation time for grinding, fine grinding and polishing of 40 min. even the
edges of the pores are sharply revealed in the SEM micrograph of Fig. 10.

In a slightly changed version this short-time diamond grinding and polishing
method is also applicable to hard metals. Sintered materials of medium or high
final porosity have to be impregnated prior to grinding and polishing. For both
sintered ceramics and structural parts manufactured by powder metallurgy processes
this pretreatment of metallographic sections is indispensable for obtaining true microstructural data by an image analysis of their polished surfaces. To avoid breaking
out of material and smearing of pores sintered iron parts were infiltrated successfully
with molten silver solder[12]. Vacuum impregnation with hot and very fluid epoxy
resins proved to be suitable for the preparation of specimens in model liquid phase
sintering experiments. Relief formation could be minimized by short polishing
times[5].

The method of revealing the microstructure with the aid of non-reactive interference layers is especially suited for the differentiation of phases in heterogeneous
and porous materials since there is no chemical or morphological alteration of the
specimen surface. Figure 11 shows a micrograph of a section through a loose

TOTAL PREPARATION TIME 24 TO 40 MIN

| STEP | PREPARATION | TIME | DISC/CLOTH | GRAIN SIZE | LOAD | R.P.M. |
|------|-------------|------|------------|------------|------|--------|
| 1 | CUTTING | | D-DISC | D 91 | | |
| 2 | COARSE GRINDING | 3 MIN | D-DISCS | D 64 | 90 N | 300 |
| 3 | | 3 MIN | | D 20 μM | | |
| 4 | FINE GRINDING | 2-3 MIN | STEEL CLOTH | D-SPRAY OR D-PASTE 3 μM | 60 N | 150 |
| 5 | POLISHING | 10-20 MIN | PELLON | | 180 N -210 N | 300 |
| 6 | | 6 MIN | SOFT SYNTHETIC | | 150 N | |

Fig. 8. A diamond grinding and polishing method for embedded sections of sintered ceramic materials.

packing of a Cu—10 w/o mixture of copper spheres and tin particles liquid phase sintered at 450°C for 15 min. The specimen was gaseous ion coated in a Bartz contrasting chamber[13,14] (Pt cathode, 1.8 kV, $O_2$ gas 0.3 mbar). With this coating the copper spheres appear orange red, the ε-phase yellow, the η-phase reaction layers green and the impregnated pores a dark green mottled tone. Vapor phase deposition of zinc selenide can be employed as well to improve the contrast between impregnated pores and different phases to such a degree that automatic structure analysis is feasible.

## APPLICATION TO SWELLING PROBLEMS OF Fe—Cu

A study of the swelling process in a Fe-10 w/o Cu liquid phase sintered model mixture furnishes evidence of the large scope of information gained by microstructural observations in powder metallurgy. Swelling is thought to be due to the penetration of melt into the grain boundaries of the solid constituents and to solid state diffusion processes. An exact analysis of the microstructural changes during liquid phase sintering seems necessary to get a deeper insight into the detailed mechanisms of the swelling process. To achieve this, compacts with a simplified starting microstructure consisting of deformed iron and copper spheres were used[1]. The Fe-Cu system was chosen because of its greater technical importance.

Figure 12 demonstrates that swelling dominates over the shrinkage process for the Fe-10 w/o Cu compacts of high initial density pressed at 589 or 785 MPa and sintered at 1165°C. A micrograph of a compact pressed at 785 MPa is shown in Fig. 13. The deformed copper particles appear in a light tone. The deformed iron spheres can be distinguished by their darker color and their grain structure. Flattened pressed contact areas have been formed between the iron particles. In this highly compacted region only small pores are observed. Generally, high density compacts consist of both regions with low porosity and strongly deformed iron particles as well as areas with higher porosity and iron particles deformed to different amounts.

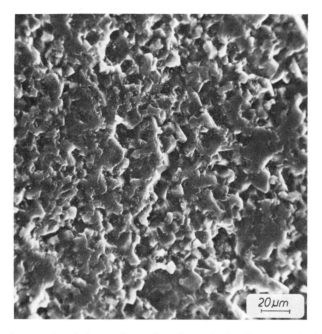

Fig. 9. SEM micrograph of the surface of a sintered alumina ceramic (99.7 w/o) after diamond cutting.

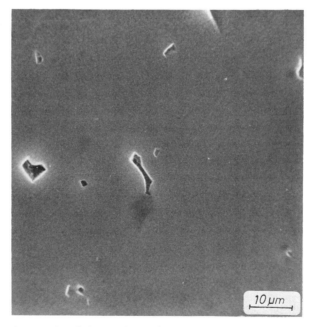

Fig. 10. SEM micrograph of the surface of a sintered alumina ceramic (99.7 w/o) after final diamond polishing. Total preparation time 40 min.

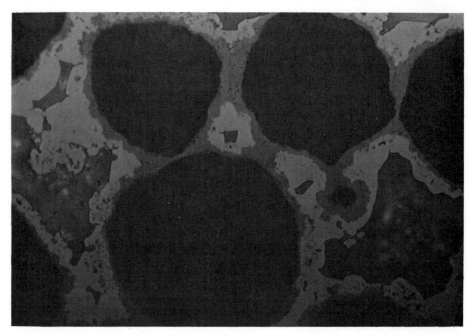

Fig. 11. Microstructure of a Cu–10 w/o Sn mixture liquid phase sintered at 450°C for 10 min.  700x.

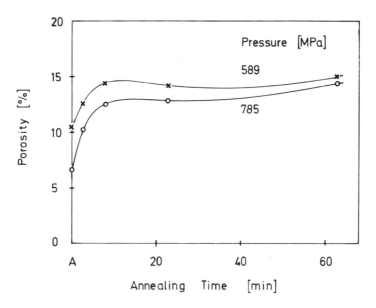

Fig. 12.  Swelling of Fe–10 w/o Cu compacts during annealing at 1165°C.

After a short annealing treatment of 5 min. at 1165°C the liquid phase penetrates completely into all pressed contact areas. Figure 14 shows an example of this type of microstructure. Gaseous ion coating (Pt; $O_2$-gas) of the polished section produces a good contrast between the iron particles (yellowish), the copper melt (orange), and the infiltrated pores (green). The evaluation of scanning electron micrographs by linear analysis yielded a mean value of 0.7 $\mu$m for the thickness of the melt layer by which the particles are separated. Competing with the rapid filling of contact areas by the melt a slower penetration of melt into the grain boundaries was observed. Figure 15 shows the corresponding microstructure after sintering at 1165°C for 8 min. The gaseous contrasting method colors the iron sphere green and the melt a yellow color. The frequency of melt-impregnated grain boundaries increases continuously up to an annealing time of 8 min. and then remains constant. The mean layer thickness of this type of film is 0.8 $\mu$m. Together with the penetration of melt into grain boundaries the originally flat interparticle contact regions are changed. As shown by the three different micrographs in Fig. 16 the straight lines of the flattened contact regions are progressively transformed into zigzag contours by annealing the compact for 3, and 8 min. at 1165°C. The peaks of this toothed pattern are in each case aligned in the direction of a melt infiltrated grain boundary of a neighboring particle. The zigzag areas of Fe-Cu solid solution appear dark after etching the polished sections by nital (Fig. 16).

Figure 17 illustrates schematically the swelling of a liquid phase sintered Fe-Cu mixture by the penetration of melt into inter-particle contact areas and into grain boundaries. Based on a thorough microstructural evaluation of polished sections, including the determination of the amounts and the chord length distribution of the different phases, a quantitative description of the co-operation of all mechanisms contributing to the observed swelling phenomena in the Fe-Cu system was possible[1]. According to Fig. 18 the total swelling of a Fe-Cu compact is explained by the superposition of four contributions relating to the penetration of melt into interparticle contact areas (I) and grain boundaries (II) and due to the volume increase of the solid Fe particles by Cu diffusion from melt films via interparticle contact zones (III) and austenitic grain boundaries (IV). For short annealing times at 1165°C there is obviously an excellent agreement between the experimental values and the calculated volume changes obtained by the addition of the microstructural contributions (I) to (IV). Deviations exist only for annealing times of 8 to 23 min. They have to be attributed to shrinkage processes. Figure 19 shows a micrograph of a specimen sintered for 23 min. at 1165°C. As a consequence of the advanced penetration of the melt along the grain boundaries single grains are separated from nearly every Fe-particle. This is the prerequisite to the so-called secondary rearrangement which has similarity to the well known primary rearrangement of solid particles at the very beginning of liquid phase sintering. Secondary grain rearrangement based on the observed particle disintegration accounts for the differences between the experimental and theoretical swelling curve. According to Fig. 18 secondary rearrangement causes a shrinkage of 1 to 2 vol %.

CONCLUSIONS

Microstructural investigations are indispensable for the proper assessment of the different stages in powder metallurgy production schedules as demonstrated by examples taken mainly from liquid phase sintering. Valuable informations on rapid dynamic processes during the course of sintering may be obtained by hot-stage scanning electron microscopy. By applying quantitative microscopy and determining the geometric properties associated with the features of metallographic sections the microstructural state of sintered materials may be quantified. Quantitative microscopic structural

Fig. 13. Microstructure of a Fe–10 w/o Cu compact pressed at 785 MPa. 750x. (Reduced 45% for reproduction.)

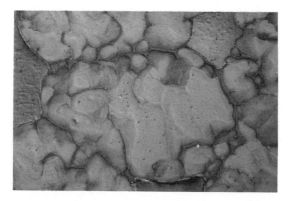

Fig. 14. Microstructure of a Fe–10 w/o Cu compact pressed at 785 MPa and annealed at 1165°C for 5 min. 800x. (Reduced 45% for reproduction.)

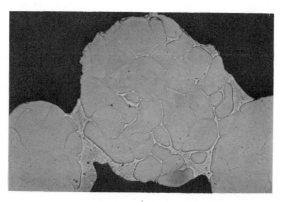

Fig. 15. Microstructure of a Fe–10 w/o Cu specimen after sintering at 1165°C for 5 min. Penetration of melt into grain boundaries of Fe particles. 700x. (Reduced 45% for reproduction.)

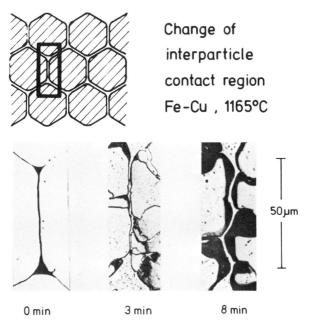

Change of
interparticle
contact region
Fe-Cu , 1165°C

50 μm

0 min          3 min          8 min

Fig. 16.  Change of interparticle contact region in a Fe—10 w/o Cu compact during sintering at 1165°C.

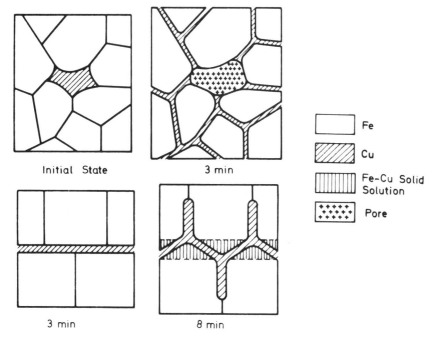

Initial State          3 min

Fe

Cu

Fe-Cu Solid
Solution

Pore

3 min          8 min

Fig. 17. Penetration of melt into interparticle contact areas (top) and into grain boundaries (bottom) during liquid phase sintering of a Fe—10 w/o Cu compact at 1165°C.

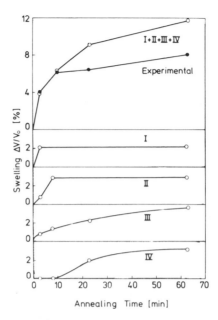

Fig. 18. Swelling of Fe—10 w/o Cu compacts pressed at 785 MPa during sintering at 1165°C. Calculated swelling contributions: I. Penetration of melt into interparticle contact areas II. Penetration of melt into Fe grain boundaries III. Cu diffusion from interparticle contact areas into Fe particles IV. Cu diffusion from grain boundaries into Fe grains.

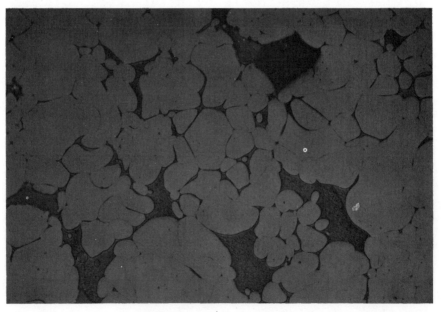

Fig. 19. Particle disintegration in a Fe—10 w/o Cu specimen after sintering for 20 min at 1165°C. 800x. (Reduced 10% for reproduction.)

analysis of powder metal parts depends, however, to a overwhelming extent on the specimen preparation techniques. Infiltration methods for porous structures, grinding and polishing of cut sections and contrasting of two- or multiphase materials must be continuously improved to satisfy the demands of automatic or semi-automatic image analysis in powder metallurgy.

## REFERENCES

1. W.A. Kaysser, "Comparative Investigations of Liquid Phase Sintering of the Systems Fe-Cu, Fe-Sn, Cu-Bi, Cu-Sn, Ni-Bi, W-Cu," *Doctoral Thesis,* Technical University, Stuttgart (1978).
2. H.E. Exner and E. Linck, "Image Analysis for Characterization of Size- and Shape Distribution of Lead Powders," *Powder Metallurgy International,* Vol. 9, p. 131–133 (1977).
3. H. Riegger, "Compacting Mechanisms during Liquid Phase Sintering," *Doctoral Thesis,* Technical University, Stuttgart (1977).
4. W.A. Kaysser, "Some Model Experiments about the Rearrangement Process during Liquid Phase Sintering," in *Contemporary Inorganic Materials 1978* (G. Petzow and W.J. Huppman, Editors) p. 41–44 (1978).
5. W.A. Kaysser, W.J. Huppmann and G. Petzow, "Comparative Metallographic Investigations of Fe-Cu, Fe-Sn, Cu-Sn, and Cu-Bi Sintered Materials," in *Pract. Metallography, Spec. Edition 10* (F. Jeglitsch, G. Petzow, Editors), Dr. Riederer-Verlag, Stuttgart, p. 553–559 (1979).
6. G. Bartz, "Contrasting of Microscopic Objects in a Gas-Ion-Reaction Chamber," *Pract. Metallography,* Vol. 10, p. 311–323 (1973).
7. D.N. Yoon and W.J. Huppmann, "The Etching Behavior of Tungsten Grains Sintered in Liquid Nickel," *Pract. Metallography,* Vol. 15, p. 399–401 (1978).
8. D.N. Yoon and W.J. Huppmann, "Shape Accommodation and Coalescence of Tungsten Grains during Liquid Phase Sintering," in *Contemporary Inorganic Materials 1978* (G. Petzow and W.J. Huppmann, Editors), p. 55–61 (1978).
9. D.N. Yoon and W.J. Huppmann, "Grain Growth and Densification during Liquid Phase Sintering of W-Ni," *Acta Met.,* Vol. 27, p. 693–698 (1979).
10. G. Elssner, S. Aldinger and S. Kuhnemann, "Problems in Preparing Sections of Brittle Materials," in *Pract. Metallography, Spec. Edition 10* (F. Jeglitsch, G. Petzow, Editors), Dr. Riederer-Verlag, Stuttgart, p. 49–60 (1979), and *Pract. Metallography* in press.
11. G. Elssner and H. Honecker, "Preparation of Microsections of Hard and Brittle Materials Using the Technotron Grinding and Polishing System," *Pract. Metallography,* Vol. 16, p. 205–214 (1979).
12. H. Metzler, D. Nitsche, "Methods of Producing Polished Metallographic Specimens of Porous Iron Green Bodies and Sintered Parts for Fully Automatic Structure Analysis," *Pract. Metallography,* Vol. 9, p. 408–411 (1972).
13. G. Petzow and H.E. Exner, "Recent Developments in Metallographic Preparation Techniques," in *Microstructural Science, Volume 3* (P.M. French, R.J. Gray and J.L. McCall, Editors), p. 291–306 (1975).
14. P. Schluter, S. Aldinger and G. Elssner, "Comparison of Phase Differentiation in Non-ferrous Metals and Composites by Gaseous Ion Coating and Chemical Etching," in *Pract. Metallography Spec. Edition 9* (W.U. Kopp and H.E. Buhler, Editors), Dr. Riederer Verlag, Stuttgart, p. 57–65 (1978).

# METALLURGICAL CONTROL OF BILLETS AND FORGINGS

J.A. Hendrickson and R.B. Sparks *

## INTRODUCTION

Wyman-Gordon Company, Eastern Division, is engaged in the manufacture of forgings, mainly for the aerospace and nuclear industries.  A wide variety of materials, including carbon, alloy, and stainless steels, aluminum, titanium, and nickel-base alloys are used.

Forging practices vary from conventional methods such as open and closed die hammer and press forgings and ring rolling, to advanced techniques involving isothermal forging and hot isostatic pressing of metal powders.

Mechanical properties and serviceability. of forgings are dependent on microstructure.  In large cross-section titanium parts, such as (shown in Fig. 1) the fan disk (700 pounds), the helicopter rotor hub (1500 pounds), the aircraft bulkhead (1300 pounds), and the 747 main landing gear support (3000 pounds), billet microstructure can be carried over into some areas of finish forged parts.  Because of this, metallurgical control of finish parts begins with the incoming billet material.  This paper will discuss the role of metallography in:
1.  Acceptance testing of incoming billet material
2.  The in-process control and final product testing of forgings
3.  Supporting investigational metallurgy

Much of the discussion will be confined to the metallurgical control of the 6Al–4V–Ti fan disk forging previously shown in Fig. 1.  The metallurgical controls necessary on this part are typical of those required on critical aerospace forgings.

## ACCEPTANCE TESTING

On billet material for the fan disk (13 inch diameter stock), slices representing the top, middle, and bottom of each ingot product are macroetched and examined for any imperfections that could be detrimental and for microstructure that would be acceptable in the finished product.

The macrostructure of fine grain 6Al–4V–Ti material acceptable as forging stock for fan disk applications is shown in Fig. 2.  Figure 3 illustrates the microstructure observed in billet slices at surface, mid-radius, and center locations.  Examples of billet microstructure which would be rejectable are those shown in Fig. 4.

*Wyman-Gordon Company, Worcester, Massachusetts  USA.

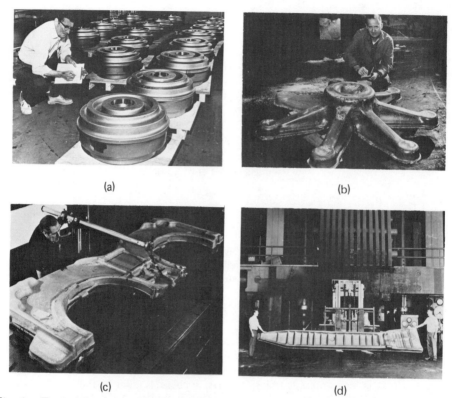

Fig. 1. Typical heavy section Ti—6Al—4V production forgings. a) Fan disc, Rolls Royce aircraft. b) Helicopter rotor hub, Sikorsky Aircraft. c) Aircraft bulkhead, McDonnell Douglas Air. d) Landing gear support beam, Boeing Aircraft.

Fig. 2. Representative macrostructure of 6Al—4V Ti billet for fan disk application. Top — transverse plane; Bottom — longitudinal plane. Magnification ∼ 1X; Etchant HF—HNO₃

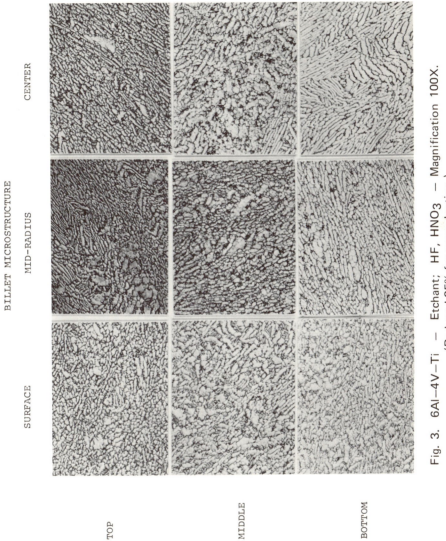

Fig. 3. 6Al–4V–Ti – Etchant; HF, HNO₃ – Magnification 100X. (Reduced 35% for reproduction.)

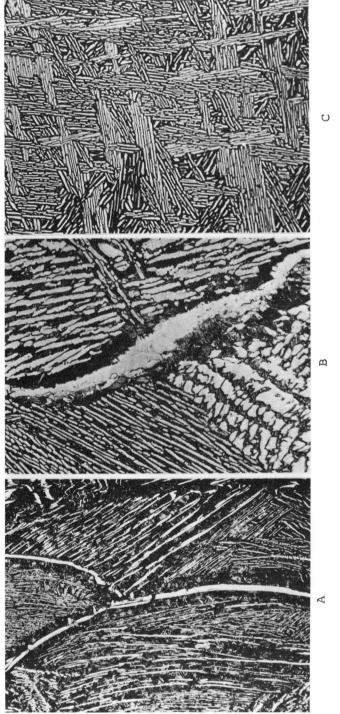

Fig. 4. Magnification 100X — Etchant: NaOH,$H_2H_2O_2$. Examples of rejectable billet microstructures: A. Stringy alpha and grain boundary outlining. B. Blocky alpha and segregation at grain boundary. C. Widmanstatten structure. (Reduced 20% for reproduction.)

The microstructure required in the fan disk forging is equiaxed or slightly elongated primary alpha in a transformed beta matrix. A uniform distribution of primary alpha between 15% and 45% is specified. The required amount of primary alpha can be obtained by adjusting forging and heat treat temperatures within a range of 50F to 75F below the beta transus.

The beta transus is that temperature at which a titanium alloy changes from the hexagonal close-packed arrangement to the body centered cubic. For the 6Al–4V–Ti alloy, this temperature is usually in the range of 1820–1830F. Slight variations in chemistry can expand this range from ~ 1775F to 1850F, and for this reason, the transus must be determined for at least the top and bottom of each heat of material. Metallographically, it can be determined by the lowest temperature at which all the spheroidal alpha goes into solution. Photomicrographs of material quenched from above and below the beta transus are shown in Fig. 5. The beta transus may also be determined by differential thermal analysis.

## PROCESS CONTROL AND PRODUCT TESTING

The morphology of the microstructure is further influenced during
1. forging
2. heat treating
3. cooling from forging and heat treating operations.
Improper processing can result in the unacceptable microstructures such as shown in Fig. 6a. If this structure is present in a finished forging, it is generally not possible to salvage or restore it to an acceptable level by further conventional processing.

Integral test material is removed from each fan disk forging after heat treatment for microstructural evaluation. Representative structures observed in forgings produced from a single heat are shown in Fig. 7. At designated intervals, forgings are destructively tested. Figure 8 shows the radial grain flow in the fan disk. Macroetched slices are used to determine if the metal is flowing properly in the die cavity, and if the grain size is uniform and of correct size. Microstructure in the various areas of the forging (Locations 1–7 of Fig. 8) is verified. Representative structures are shown in Fig. 9.

In some instances, when processing techniques are suspect, or customer requirements are for final microstructure verification directly on the forgings, this is accomplished by examination with portable polishing and metallographic equipment such as is shown in Fig. 10. Closed circuit TV may be used for group observation and microstructures are documented either by replicas of the surfaces or direct photography. In some instances, customers require that the replicas and/or photomicrographs be submitted with the other quality control documentation as part of the certification package.

## INVESTIGATIONAL METALLOGRAPHY

In addition to routine examination of starting billet material and finished forgings, an important function of metallography in the forging industry is investigational. These investigations include evaluation of abnormalities detected during acceptance testing, such as is shown in Figs. 11 and 12, segregation of Laves and $Ni_3Cb$ phase in Inco 718 Ni-base material, Fig. 13, hard alpha segregate in 6Al–4V–Ti, and Fig. 14, beta flecks or alpha nude areas in 6Al–4V–Ti.

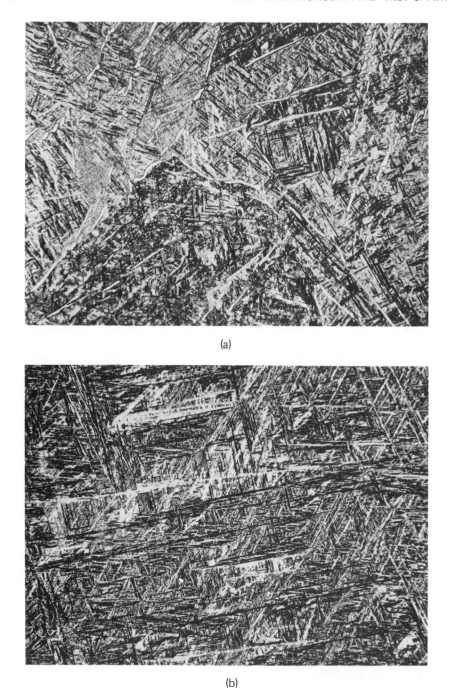

(a)

(b)

Fig. 5.  Beta transus determination sample.  a) Water quenched from beta transus minus 20F.  Primary alpha and alpha prime.  b) Alpha prime and prior beta grain boundaries water quenched from above beta transus.  Magnification 200X — Etchant: NaOH–$H_2O_2$

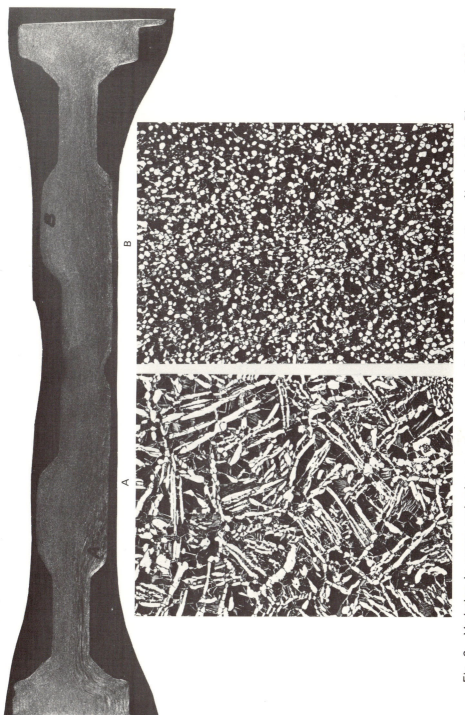

Fig. 6. Variation in macro and microstructure as a result of a furnace malfunction. A) rejectable   B) acceptable  Photomicrographs 100X; macrograph approximately ½X. (Reduced 15% for reproduction.)

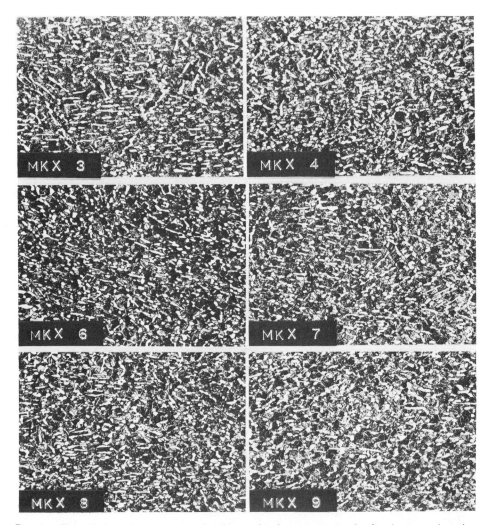

Fig. 7. This shows the representative integral microstructures in forgings produced from the same heat. Magnification 100X. Etchant: NaOH–$H_2O_2$. (Reduced 30% for reproduction.)

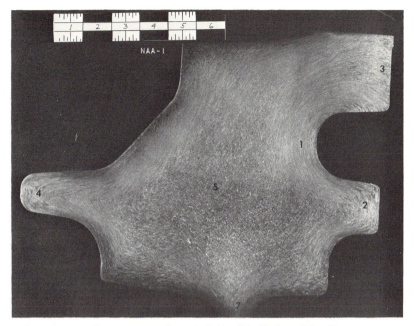

Fig. 8. Radial grain flow section from a 6Al—4V—Ti fan disk. Microstructure is veri-fied in locations 1 through 7.

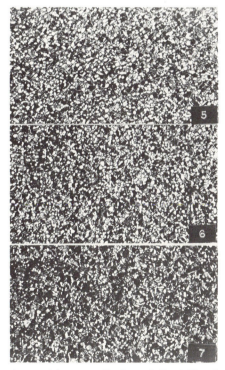

Fig. 9. Microstructures observed in areas 5, 6, and 7 of Fig. 8. Magnification 100X
Etchant: NaOH—$H_2O_2$   (Reduced 50% for reproduction.)

Fig. 10. Top: Large Ti forging being electropolished for direct examination of microstructure. Bottom: Rating ASTM grain size on Waspaloy forgings.

Fig. 11. Segregation indicated by arrows. Etchant: HF, $HNO_3$, HCL. Magnification 1X. Material — Inco 718    (Reduced 60% for reproduction.)

Fig. 12. A. Light photomicrograph showing segregate of Ni$_3$Cb needles and Laves phase (Magnification 100X). B. SEM photograph at 500X. C. X-ray map showing Cb enrichment. D. Microstructure away from segregate.

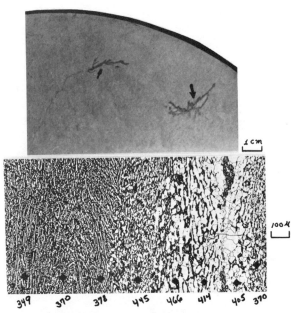

Fig. 13. Material 6Al−4V−Ti showing: Top: Macroetch indication indicated by arrows. Bottom: Light photomicrograph showing hard alpha segregation probably due to oxygen stabilization. Etch: HF−HNO₃, ammonium bifluoride showing alpha enriched area and micro hardness survey (500 gram load).

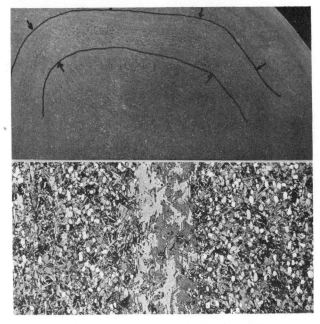

Fig. 14. 6 Al−4V−Ti alloy. Top: 1X showing macro indications. Bottom: 100X showing beta flecks. (Reduced 50% for reproduction.)

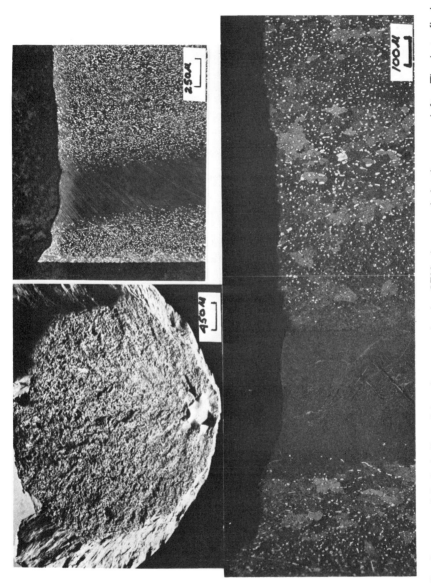

Fig. 15. The fracture origin is indicated by the arrow in the SEM photograph in the upper left. The beta fleck is visible at the fracture origin in the two light photomicrographs. Alloy: 6Al–6V–2Sn. Problem: low ductility. Cause: beta fleck.

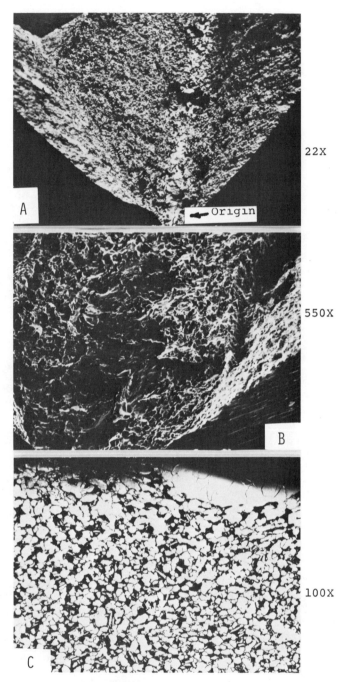

Fig. 16. Alloy: 6Al–4V–Ti. Problem: low ductility. Cause: alpha segregation. (A&B) SEM of fracture origin. (C) Light photomicrograph showing alpha segregation. Etch: NaOH, $H_2O_2$. (Reduced 20% for reproduction.)

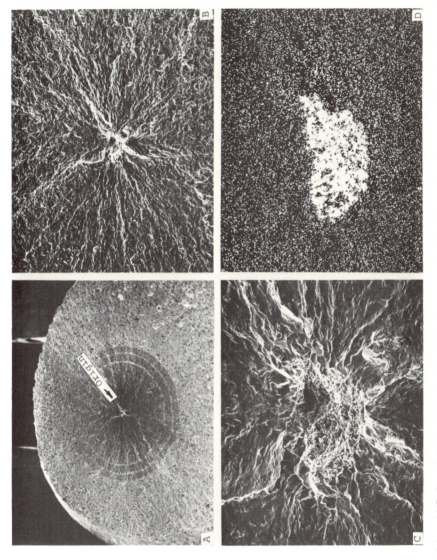

Fig. 17. Material — HIP Rene' 95 powder showing aluminum contamination at the fracture origin of a low cycle fatigue test. A. 22X secondary electrons. B. 110X secondary electrons. C. 550X secondary electrons. D. 550X Al X-ray map. (Reduced 30% for reproduction.)

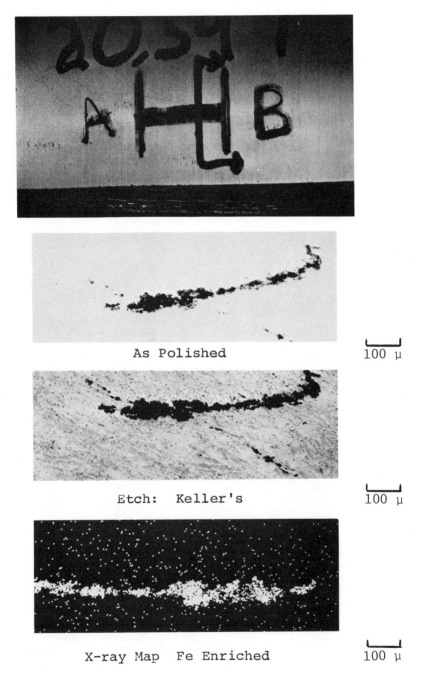

As Polished                                    100 μ

Etch:   Keller's                               100 μ

X-ray Map   Fe Enriched                        100 μ

Fig. 18. X-ray map Fe enriched. Alloy: 7075 Al. Problem: Subsurface ultrasonic indication elongated from A to B. Cause: Fe rich inclusion probably due to scrap contamination.

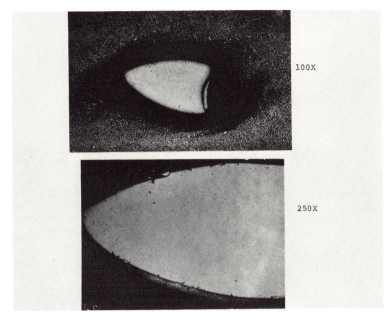

Fig. 19. Material 6Al—4V—Ti showing tungsten inclusion detected by radiography. (Reduced 40% for reproduction.)

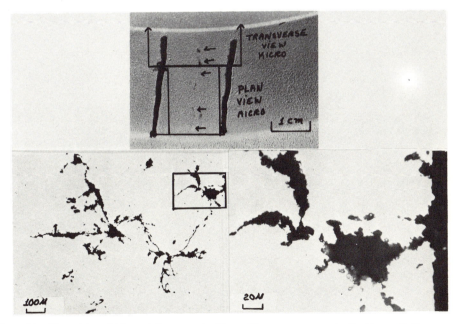

Fig. 20. Material — 316 stainless steel. Top: Dye penetrant indication. Bottom: Voids and nonmetallic inclusions revealed by microexamination.

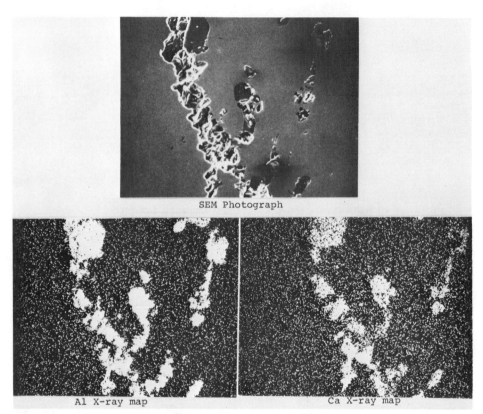

SEM Photograph

Al X-ray map                                          Ca X-ray map

Fig. 21.  This shows an inclusion which is enriched in Al and Ca.  Mag. 1000X.
(Reduced 35% for reproduction.)

Examples of mechanical test failures associated with material segregation and/or contamination are shown in Fig. 15, beta fleck associated with low ductility in 6–6–2–Ti, Fig. 16, alpha segregation also associated with low ductility in 6Al–4V–Ti, and Fig. 17, aluminum contamination in HIP Rene' 95 Ni-base powder material.

Identification of indications detected by ultrasonic testing, radiography, penetrant and magnetic inspection is shown in Figs. 18 through 21. Forgings rejected by customers and analyses of failed parts are also important to the overall quality control task.

In summary, the final quality of forgings is directly dependent on microstructure, and each of the following processes contribute to this final structure:

1. ingot melting
2. ingot to billet conversion
3. forging and heat treating

Strict microstructural control must be maintained in all these areas to provide satisfactory forgings which will meet the customers' specifications.

# METALLOGRAPHIC QUALITY CONTROL OF WELDING AND BRAZING

G.M. Slaughter*

## INTRODUCTION

Metallography serves as an invaluable quality control technique in many welding and brazing operations in the energy field. It often supplements nondestructive testing processes such as dye-penetrant and radiography. Additionally, it can provide information ordinarily not obtainable by conventional inspection procedures. This paper describes a number of case histories illustrating the value of metallography in assuring integrity in the fabrication of metals and components in energy systems.

## MEASUREMENT OF FERRITE IN STAINLESS STEEL WELDS

To minimize the possibility of hot cracking during welding of the austenitic stainless steels, it is common practice to produce welding filler metals of such composition that the weld deposit contains a small amount (on the order of five percent) of delta ferrite. Both crack susceptibility and mechanical behavior are related to the amount and morphology of ferrite in the weld metal. It thus often becomes important to measure the amount and distribution of ferrite in representative weld deposits[1]. Metallography provides additional information to that obtained from other ferrite-measuring techniques.

Figure 1 illustrates the extent of cracking which can occur when inadequate ferrite control is maintained. Figure 2 allows a comparison of ferrite distribution in welds produced by different processes. It appears that the "coarseness" of the ferrite distribution depends primarily upon the substructure size, that is, essentially upon the relative weld energy input utilized. The ferrite morphology is thus quite process dependent.

## BRAZING OF THERMOCOUPLES TO TEST COMPONENTS

Metallographic examination provides an excellent means for assessing the reliability of brazed joints including those which secure thermocouples to test components. A photomicrograph of a satisfactory braze of a stainless steel-clad thermocouple to the inside of a fuel rod sheath is shown in Fig. 3. A joint with insufficient brazing filler metal is shown in Fig. 4. The dry-hydrogen furnace brazing procedures were modified sufficiently to produce consistently reliable brazed joints of this type.

*Metals and Ceramics Division, Oak Ridge National Laboratory, Oak Ridge, Tennessee 37830 USA.

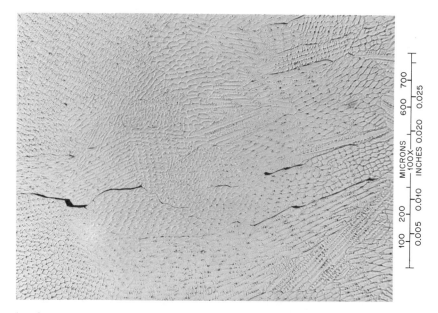

Fig. 1. Cracking in stainless steel weld metal resulting from an inadequate amount of delta-ferrite.

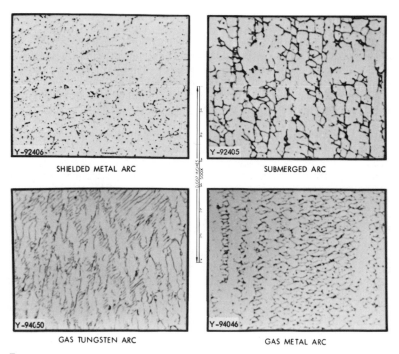

SHIELDED METAL ARC                SUBMERGED ARC

GAS TUNGSTEN ARC                  GAS METAL ARC

Fig. 2. Ferrite distribution in stainless steel weld metal deposited by different welding processes.

Fig. 3. Stainless steel-sheathed thermocouple brazed to the inside of a stainless steel fuel rod sheath. Good filleting and wetting is observed.

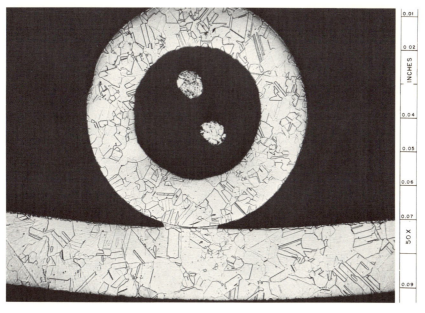

Fig. 4. Thermocouple-to-sheath braze with an inadequate amount of brazing filler metal.

CRACKING AND POROSITY IN REFRACTORY METAL WELDS

The high melting points and high strengths of refractory metal alloys at elevated temperatures make them extremely attractive as structural materials for certain energy systems[2,3,4]. On the other hand, these alloys have low tolerance for impurities during fabrication and welding. Impurity contamination can result in increased brittleness of the weldment and cracking and porosity. The large grain size usually associated with refractory metal weld deposits also is an important factor in raising their ductile-to-brittle transition temperatures (increased brittleness). Metallography serves as an essential tool for evaluation of these various factors in weld reliability.

Microcracks which developed upon welding of highly contaminated (4200 ppm oxygen) Nb—1 wt % Zr alloy are shown in Fig. 5. Tungsten is particularly subject to cracking during welding, even if extreme precautions are taken to avoid impurity pickup and excessive restraint of the joint. A sound gas tungsten-arc weld is shown in Fig. 6; severe cracking in a similar weld is shown in Fig. 7.

BERYLLIUM

The welding of end caps in beryllium tubes is a difficult technical problem[5]. Residual volatile constituents in the base metal lead to weld metal porosity. There is also a general tendency of veryllium to crack during welding. It has been found that the general weldability of beryllium is greatly dependent upon the degree of purification of the base metal obtained during the fabrication process. Metallography serves as the best means for evaluating the weld metal soundness associated with each product.

Fig. 5. Microcracks which may develop upon welding highly contaminated Nb—1 wt % Zr Alloy.

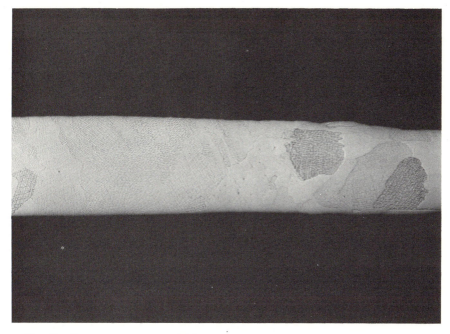

Fig. 6. Gas tungsten-arc weld in tungsten sheet. No defects are evident.

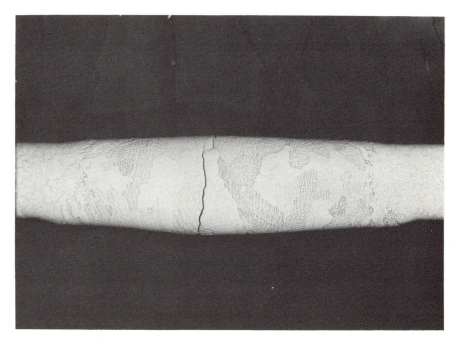

Fig. 7. Tungsten weld with intergranular cracking.

Figure 8 is a gas tungsten-arc weld exhibiting extensive porosity, while root cracking is in evidence in Fig. 9.

## TUBE—TO—TUBESHEET JOINTS

Tube-to-tubesheet joints for heat exchangers are notably difficult to inspect by conventional nondestructive techniques. Thus, metallography serves as a very useful means for developing optimized fabrication procedures. An Inconel weld with severe cracking in the weld resulting from improper welding technique is shown in Fig. 10. Figure 11 illustrates a large pore which is virtually undetectable by conventional non-destructive procedures.

## TUBE—TO—FIN BRAZING

Metallography serves as an essential means for evaluating many aspects of brazed joints. One application of brazing, the attachment of stainless steel-clad-copper high-conductivity fins to stainless steel tubes, is particularly dependent upon metallography [6]. During high-temperature service, uncovered copper edges permit extensive oxidation of the copper (Fig. 12). Acceptable bonding but poor edge coverage is shown by metallographic sectioning (Fig. 13). By use of an optimized amount of brazing filler metal, the excellent bonding and edge covering shown in Fig. 14 can be obtained.

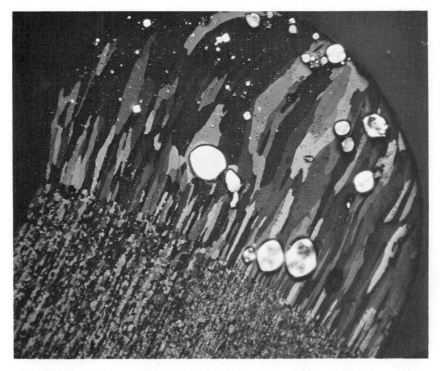

Fig. 8. End-closure weld in beryllium tubing exhibiting extreme porosity.

Fig. 9.  End-closure weld in beryllium tubing exhibiting root cracking.

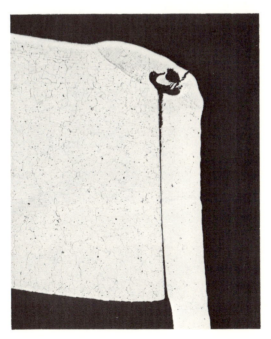

Fig. 10.  Severe cracking in an Inconel tube-to-tubesheet weld resulting from improper welding techniques.

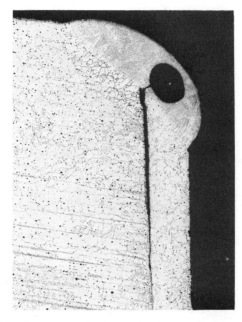

Fig. 11.  Large pore in Inconel tube-to-tubesheet weld.

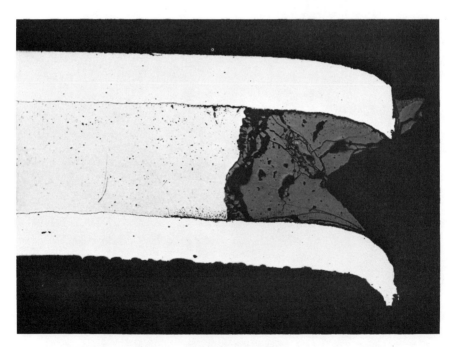

Fig. 12.  Severe high-temperature oxidation of copper in stainless steel-clad-copper
high-conductivity fin resulting from unprotected edge.

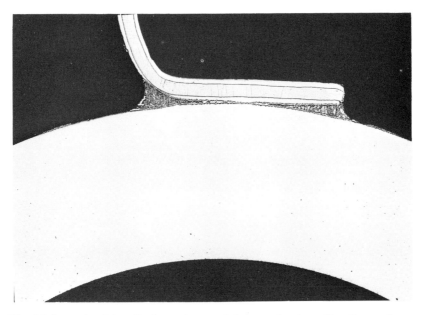

Fig. 13. High-conductivity fin brazed to stainless steel rube. Excellent adherence is evident but the copper edge is not protected.

Fig. 14. Brazed high-conductivity fin showing edge protection of copper.

DISSIMILAR METAL WELDS

A problem of great current interest in the production of electric power from fossil-fired plants involves the failures of dissimilar metal welds in superheater and reheater sections[7,8]. Most superheaters and many reheaters delivering steam at temperatures of $1000°F(538°C)$ or more contain austenitic stainless steel tubing, whereas the tubing in the lower temperature regions is made of ferritic steels. An increasing number of failures in the transition joints connecting these two materials has occurred in recent years.

Figure 15 illustrates a crack in the heat-affected zone of the 2¼ Cr−1 Mo ferritic steel. A higher magnification view showing the nature of the crack tip is presented in Fig. 16. The Steam Power Panel of the Joint Committee of ASTM, ASME, and MPC) has formed a Task Group to assess the impact of these failures and to formulate a plan to alleviate the problem.

SUMMARY

From the above-mentioned case histories, it is evident that metallography plans an integral role in the quality control of welded and brazed joints. As additional advances are made, it should become an even more important tool.

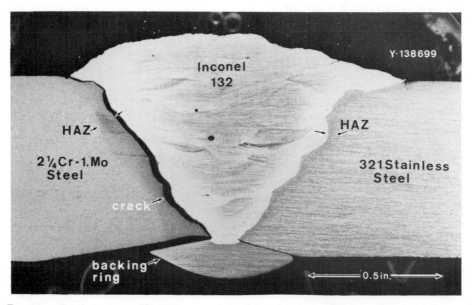

Fig. 15. Crack in heat-affected zone of ferritic steel immediately adjacent to Inconel weld metal.

Fig. 16.  View of crack tip in failed dissimilar metal weld transition joint.

REFERENCES

1.  G.M. Goodwin, N.C. Cole, and G.M. Slaughter, "A Study of Ferrite Morphology in Austenitic Stainless Steel Weldments," *Weld J. (Miami)* 51(9): 425-s-429-s (September 1972).
2.  A.J. Moorhead and G.M. Slaughter, "Welding Studies on Arc-Cast Molybdenum," *Weld. J. (Miami)* 53(5): 185-s-191-s (May 1974).
3.  E.A. Franco-Ferreira and G.M. Slaughter, "Welding of Columbium-1% Zirconium," *Welding J. (N.Y.)* 42(1), 29—36 (January 1963).
4.  N.C. Cole, R.G. Gilliand, and G.M. Slaughter, "Weldability of Tungsten and Its Alloys," *Welding J. (N.Y.)* 50(9), 419-s-426-s (September 1971).
5.  R.G. Gilliland and G.M. Slaughter, "Fusion Welding of End Caps in Beryllium Tubes," *Welding J. (N.Y.)* 42(1), 29—36 (January 1963).
6.  G.M. Slaughter, E.A. Franco-Ferreira, and P. Patriarca, "Welding and Brazing of High Temperature Radiators and Heat Exchangers," *Welding J. (N.Y.)* 47(1), 15—22 (January 1968).
7.  R.J. Gray, J.F. King, J.M. Leitnaker, and G.M. Slaughter, *Examination of a Failed Transition Weld Joint and the Associated Base Metals,* ORNL—5223 (January 1977).
8.  J.F. King, M.D. Sullivan, and G.M. Slaughter, "Development of an Improved Stainless Steel to Ferritic Steel Transition Joint," *Weld. J. (Miami)* 56(11): 354-s-358-s (November 1977).

# METALLOGRAPHIC CONTROL OF AEROSPACE COMPONENTS

J.W. Hutchinson, R.K. McLeod, T.W. Heaslip*

## INTRODUCTION

There have been many tragic air disasters occur in the aviation industry which, because of their nature, were highly publicized and created a strong public awareness of the need for aviation safety. "A chain is only as strong as its weakest link"; and so it goes for aircraft and their components. Aviation safety involves the man, the machine, and the environment. The machine is controlled by its components and effective control of those components will ensure the safety of the machine.

Metallographic control of aerospace components is a broad term and is used during the design stage, production and in some respects maintenance and overhaul. Metallography is used extensively as a tool in failure analysis to help determine cause of failure and subsequently as a control method to help establish where the control sequence broke down, hopefully preventing a recurrence.

## CASE HISTORIES

Two examples are presented which investigate the failure of an aerospace component and demonstrate how metallography may have been used more effectively as a quality control tool to detect flaws prior to service. Two additional examples are presented which show how metallography was used in failure analysis to detect a breakdown in control which occurred during the maintenance and overhaul stages.

### AIRCRAFT WHEEL FAILURE (Magnesium Alloy Casting, AZ91C–T6)

While taxiing for take-off on a prepared gravel strip, the left main wheel on a twin engine aircraft failed and a portion of the flange was projected away from the wheel assembly. The failed sections of the wheel were recovered (Fig. 1) and the preliminary visual examination disclosed evidence of a pre-existing crack as indicated by local discoloration along the flange fracture adjacent to the wheel rim tire bead seating area (Fig. 2). A failure analysis was conducted on the wheel to determine the cause of failure. It can be concluded from the results of this analysis that metallographic quality control of this component during manufacture could have prevented the failure.

Sections of the wheel flange which contained the pre-crack area were examined in the Scanning Electron Microscope (SEM). The total fracture area consisted of the pre-crack zone and the final fracture zone, both of which were overload in nature.

*Aviation Safety Engineering Facility, Aviation Safety Bureau, Transport Canada, Ottawa, Ontario K1A ON8, Canada.

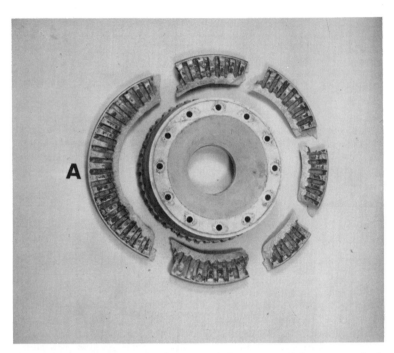

Fig. 1.  Failed wheel as recovered.  Portion of flange containing pre-crack A.

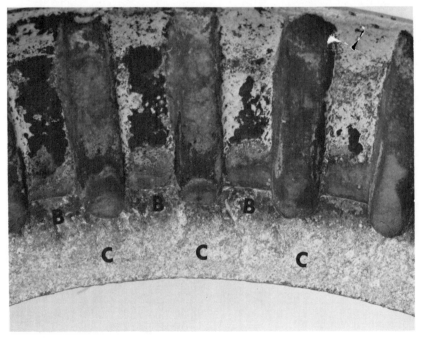

Fig. 2.  Discolored pre-crack region B and clean fracture area C.

SEM examination of the pre-cracked zone revealed numerous areas of nodular shaped clusters, separated by a semi-continuous network of voids (Fig. 3). This pattern is characteristic of shrinkage porosity in a cast material and is an original casting defect. The production radiographs for the wheel only revealed a barely discernible presence of shrinkage porosity, even though knowledge of its exact location was now known. It was concluded that the radiographs did not show any clear evidence of casting defects which would cause rejection of the wheel.

SEM examination of the final overload zone revealed a fracture surface consisting of a blocky lamellar phase as well as a smooth glassy phase containing many brittle fissures (Fig. 4). SEM examination of a laboratory induced overload of another section of the wheel rim revealed the same fracture characteristics as the in-service overload. Standard tensile tests were conducted on samples taken from the rim of the wheel and the results showed an unusually high yield strength, but also showed a drastic reduction in ultimate tensile strength and ductility of the failed wheel as compared with typical AZ91C—T6 properties. The test results marginally exceeded minimum specification requirements.

A sample of the casting was subjected to wet chemical analysis. The material met all the requirements of AZ91C but one, that being the presence of an excess of aluminum.

Sections were cut perpendicular to the fracture surface for metallurgical examination (Fig. 5). The most marked feature was fissuring evident in the massive (white) grain boundary phase. This white grain boundary phase was determined by x-ray spectrometric analysis to be a magnesium/aluminum/zinc intermetallic phase and compares with the smooth glassy fissured phase observed in the SEM. In the T6 heat treated condition this phase should be dispersed as a fine lamellar precipitate throughout the structure, similar to the islands of lamellae in the surrounding areas. The massive islands between the lamellar phased areas show some evidence of a fine Widmanstatten precipitate pattern. The structure indicates the casting was solution heat treated and artificially aged, but given the excess aluminum content, the solution treatment was probably followed by too slow a cooling rate to maintain the alloying elements in solid solution.

It was concluded from the failure analysis of the wheel that the failure occurred during a momentary high stress condition at a section that had been weakened by an original casting defect. The failure was aided by the fact that the material was of unusually low strength and ductility as a result of improper alloy composition and the resulting inadequate heat treatment.

The quality control of this type of wheel was through the use of production radiographs. Metallographic examination of selected samples from the production batch would have been a suitable back-up quality control method and in this case revealed a cause for rejection, thus preventing a failure.

GEAR FAILURE (Low Alloy Steel)

While in level flight at 800 feet above ground the gas turbine engine stopped on the single engine helicopter. During the emergency landing the helicopter suffered major damage. Tear down of the engine revealed that the idler bevel gear in the accessory drive assembly had failed and jammed the gears (Fig. 6).

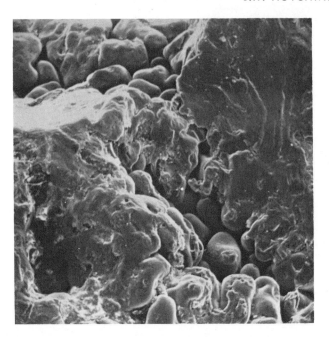

Fig. 3. Pre-crack surface displaying nodular shaped clusters and semi-continuous network of voids characteristic of shrinkage porosity. (X400).  (Reduced 10% for reproduction.)

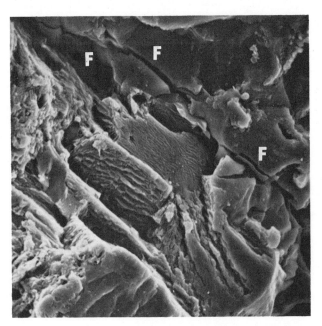

Fig. 4. Final overload surface shows blocky lamellar phase and brittle fissured phase F. (X2000).  (Reduced 10% for reproduction.)

Fig. 5.  Fissuring evident in the massive (white) grain boundary phase.  (Nital etch, X200).

Fig. 6.  Accessory drive gear assembly shows failed gear jamming the system.  (½ size approximately).

The bevel gear had fractured into three sections (Fig. 7). Preliminary optical examination indicated that fatigue fracture had originated at the roots of at least two teeth and propagated until complete fracture of the section occurred. SEM examination confirmed the fracture mode as low stress/high cycle fatigue growth from the corner at the tooth root radius (Fig. 8). There was some slight post-fracture damage observed in the fatigue origin area but no adverse mechanical or geometrical stress concentrations were observed.

The gear conformed to dimensional specifications and X-ray spectrometric qualitative analysis of the gear material did not reveal any apparent discrepancies from the specified AMS 6260H (SAE 9310) material. Samples were cut from the gear at various locations for metallurgical analysis. Polished and etched microsections displaying the gear tooth profile revealed a gross metallurgical discrepancy in the gear. The carburized case hardened layer was non-existent at the root fillets of each tooth (Fig. 9). The absence of the case hardened layer had lowered the resistance to bending fatigue and created a severe metallurgical stress concentration at the root fillets. Fatigue cracking subsequently initiated under normal loads and failure of the gear resulted.

The manufacturing specifications called for a finish grinding of the gear after the carburizing heat treatment and the amount of material removal was based on the case depth before finishing. In the subject gear the depth of case was not great enough and thus the case was removed at the tooth roots during the finish grinding process.

Fig. 7. Failed gear with fatigue origins $F_1$ and $F_2$ indicated. Large arrow indicates direction of rotation.

Fig. 8.  Low stress/high cycle fatigue growth from corner of tooth root radius. (X20).

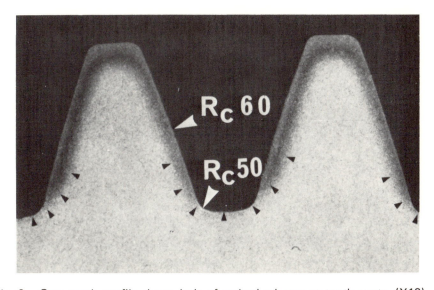

Fig. 9.  Gear tooth profile shows lack of carburized case at tooth roots. (X10).

Metallurgical quality control of the finished gear is achieved through magnetic particle inspection, temper etching, hardness tests and occasionally, destructive analysis. Metallographic analysis of samples or test coupons would have established the lack of case in the subject gear.

## CYLINDER FAILURE (Low Alloy Steel)

During cruise flight a light single engine aircraft lost power which resulted in a forced landing. Investigation revealed that one of the cylinder barrels had split circumferentially and separated into two pieces (Fig. 10). Optical and SEM examination confirmed the mode of failure as fatigue which had progressed around 50% of the barrel circumference before total cylinder failure occurred.

The cylinder barrel bore was chrome plated, a process used as a salvage operation at remanufacture. The cylinder bore is smoothly honed prior to plating with porous chromium and then electrochemically etched to produce a network of microcracks which is ttermed channel porosity. This is considered advantageous in wear applications in which lubrication is required, because it promotes wetting action and provides oil retention after initial lubrication. The etched surface is then finished by honing or lapping.

Sectioning of the cylinder barrel revealed the chrome layer and network of cracking (Fig. 11). In the vicinity of the fatigue fracture origin a mechanical notch in the bore wall was observed (Fig. 12). The chrome plating was tightly adherent to the notch, showing that this defect was present prior to plating. The notch was a severe mechanical stress concentration which resulted in fatigue initiation under normal service loads.

In the subject analysis, metallography was used to detect a flaw which caused a failure and established that the flaw was introduced during the overhaul stage. As a quality control item, this analysis illustrates that metallography may be used after the fact to help improve control.

## PROPELLER PILOT TUBE (Low Alloy Steel)

While an engine run-up was being conducted on a twin engine aircraft, the left propeller pilot tube failed causing the propeller to go into fine pitch and overspeed. The tube was found separated at the threaded end adjacent to the "O" ring groove and the locking nut.

Optical (Fig. 13) and SEM (Fig. 14) examination of the fracture surface disclosed three distinct pre-crack regions displaying on intergranular, low-energy mode of cracking. X-ray spectrometric analysis of the pre-cracked surfaces revealed strong indications of braze metal constituents (copper, zinc, nickel) which were in addition to the low alloy steel base material.

Metallurgical sections taken from an area adjacent to the fracture showed that the tube had been reworked in this area by brazing (Fig. 15). The microstructure on either side of the fracture was representative of a heat affected zone centered around the "O" ring groove where the braze reworking was performed. Hardness tests confirmed the heat affected zone and revealed a range of hardnesses varying from Rockwell C 21 to Rockwell C 53. Specifications for the part required a range of Rockwell C 30–38.

Fig. 10.  Cylinder barrel split in two due to fatigue cracking and subsequent over-load.

Fig. 11.  Cylinder barrel bore surface shows chrome layer and network of cracking. (X100).

Fig. 12. Cylinder barrel bore shows mechanical notch at fatigue origin. (X125).

Fig. 13. Fracture surface showing pre-crack regions P.

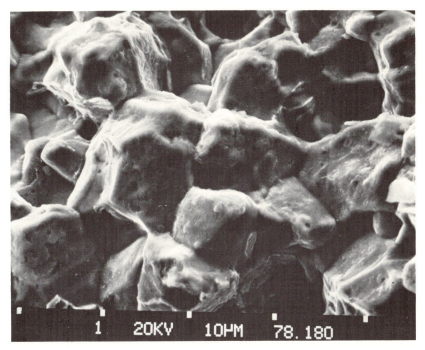

Fig. 14. Pre-crack surface displays intergranular, low-energy mode of fracture. (X2440).

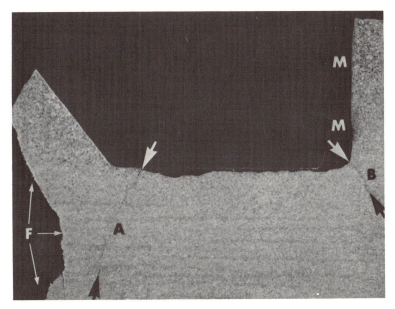

Fig. 15. Section through "O" ring groove. Braze metal M, fracture surface F, secondary cracks A and B.

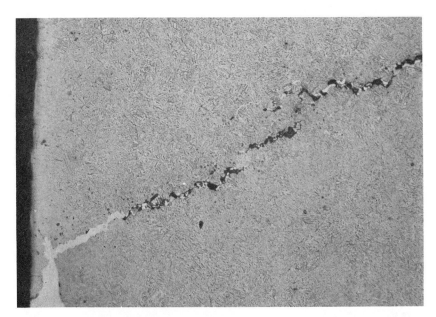

Fig. 16.  Liquid braze metal penetration in secondary crack.     (X200).

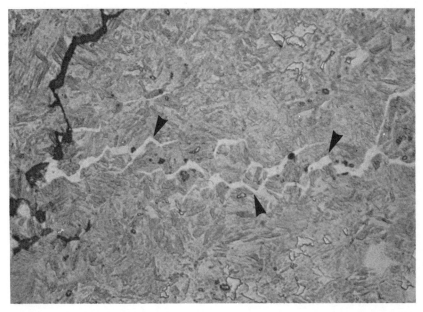

Fig. 17.  Liquid braze metal penetration in small fissures near secondary crack tip.
(X1000).

Secondary cracking was also observed beneath the braze metal. Liquid braze metal penetration was evident at the mouth of the cracks (Fig. 16) as well as in small fissures near the crack tip (Fig. 17). The mode of fracture of the pre-cracks, supported by features observed in the secondary cracks, is characteristic of liquid-metal embrittlement by the braze metal which contains copper and zinc, known embrittling metals.

Braze reworking of the part occurred during field maintenance. It is not advisable and replacement is recommended by the manufacturer. Failure analysis of the tube illustrates again how metallography can be used to pin point a source of failure (in this case field maintenance) and thus help to improve control of the component.

## SUMMARY

This paper has illustrated only a few examples of how metallography may be used for quality control of aerospace components. The examples chosen have each involved the failure of a specific part. Examining the process of quality control from this "Failure Analysis" viewpoint allows the engineer to establish more effective control of the component by indicating where it may be lacking.

.Quality control does not end in the production stage, it exists throughout the life of the part. The failure analysis examples have used metallography to show how quality control broke down in the production stage, as well as in the overhaul and maintenance stages. Failure analysis can be used to improve quality control through metallography; by using it as a tool where it is lacking, by improving the techniques where it is used and for establishing good maintenance and overhaul practices. In this way aviation safety can also be improved.

# METALLOGRAPHY OF ARMAMENT HARDWARE

L.E. Samuels*

## INTRODUCTION

It is one of the ironies of metallurgy that, throughout history, the most-advanced and highest-quality metals have been used largely in armaments, both offensive and defensive.  Sword-making was one of the first highly-developed metallurgical technologies, followed by that of the production of body armour.  Quality control has also been of paramount importance in armament manufacture because of the need for reliable and safe operation under arduous and dangerous conditions.  Moreover, these equipments often are subjected to stress conditions that cannot be analysed precisely and under which the behaviour of metals is not well understood.  Consequently, the tendency is to play safe and demand high quality standards.

These days, armaments vary enormously in their metallurgical complexity and sophistication and it is not possible to consider them all here.  The present discussion will be confined to the simpler end of the spectrum of sophistication;  specifically, to conventional medium-calibre guns and their ammunition.  The emphasis, moreover, will be on aspects of quality control that are of unique importance in these applications.  Although metallography is used commonly in production quality control of armament hardware of this nature, probably a more important role of metallography is in establishing, by investigations of problems that arise in production and use, the special quality standards that need to be applied.  It will be most profitable to concentrate on the latter aspects here.

## SOME GENERAL CHARACTERISTICS OF GUNS AND AMMUNITION

The essential operative unit of a gun is a thick-walled pressure vessel (called a *gun tube* or *gun barrel*) which may either be a monoblock or separated into a jacket and a replaceable liner tube (Fig. 1a).  In either event the tube is made from a forging of a high-strength quenched-and-tempered steel which is bored out to the calibre required.  Helical grooves are machined into the bore of the tube, the purpose of this *rifling* being to impart rotation to the projectile and so ensure stability in flight.  The bore is also machined at the breech end to a chamber of suitable shape to accept the projectile and the charge of propellant.  A heavy forged-steel end fitting (the *breech ring*) is attached to the breech end of the tube, this fitting containing an aperture, sealed with the *breech block*, through which the projectile is inserted.  Gun tubes are subjected to unusual service conditions of very high pressures, erosion by the hot gases produced when the propellant burns, and wear by the projectile.

---

*Materials Research Laboratories, Department of Defence, Melbourne, AUSTRALIA.

A projectile may be solid or contain a cavity that is filled with high explosive. In the latter case, the round is called a *shell.* Shells of larger calibres commonly are machined from hot-pierced, hot-drawn steel forgings, but shells of smaller calibres may be either machined from solid bar or deep drawn from plate. A thread of high-quality may have to be machined in the mouth of the body to accept the *fuze* that detonates the explosive filling on arrival at or near the target. A groove is formed towards the base of the body into which a ring of soft material (called the *rotating band* or *driving band*) is fitted; the band most commonly is made of a copper alloy but may be a plastic. The purpose of the rotating band is to engage in the rifling grooves to provide a seal to the propellant gases and to impart rotation to the round as it proceeds up the bore of the gun tube. The relationship between these components is illustrated in Fig. 1b.

The propellant charge often is contained in a deep-drawn metal can, called a *cartridge case,* which has thin walls and a thick base into which is fitted devices to initiate burning of the propellant (Fig. 1b). The cartridge case may be loaded into the gun chamber as a separate unit or it may be attached to the shell body, as indicated in the sketch in Fig. 1b.

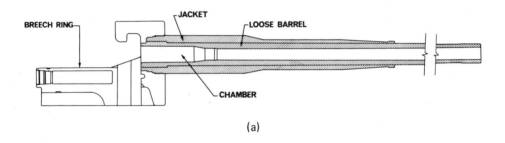

(a)

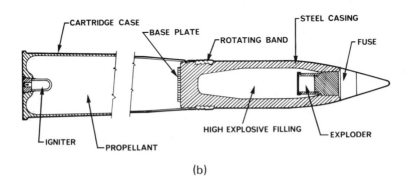

(b)

Fig. 1. Sketches illustrating the significant features of a medium-calibre naval gun tube (a), and a round of high-explosive ammunition (b). The two sketches are, of course, not to the same scale.

Numerous variations of the above are now found.  Moreover, numerous other critical metal components are present in the gun mounting, the shell loading and extraction mechanism, the shell fuze, and the igniter.  It will suffice, however, to confine the present discussion to gun tubes, shell and cartridge cases described generally above as characterizing the special problems that arise with items of armament hardware that have to operate in association with explosives, this being the unique feature of the subject.

## GUN TUBES

The bore surface of a gun tube is scoured by hot gases of complex composition each time that a round is fired, the conditions being most severe at the breech end.  This has three major consequences.  First, marked erosion occurs gradually enlarging the bore towards the breech end (Fig. 2).  This effect can be, and once always was, the one that set the limit to the useful life of the tube.  Secondly, thermal stresses produce a network of cracks in the bore surface (Fig. 3).  These cracks are produced during the firing of the first round or two and extend with further firings.  They soon reach a rather open form and typically extend for about 1mm (0.04in) in depth to thermal effects alone (crack at left in Fig. 4).  Thirdly, the microstructure of the surface layers of the bore is altered (Fig. 4).  The structure of a comparatively deep layer is affected essentially by thermal effects alone (Fig. 4), but the chemical composition of a shallower surface layer is also altered;  the latter layer is recognizable in optical microscopy (although it is not resolved in Fig. 4) as a white-etching layer and may appear to be comprised of two distinct inner and outer layers.  This layer of altered microstructure obviously is of considerable metallographic interest but currently its microstructure and composition are poorly understood[1].  In addition to these three effects due to the propellant gases, a layer of copper alloy from the rotating band is deposited on the bore surface and worked into the thermal cracks.

Fig. 2.  Bore surface at the chamber end of a gun tube in which many rounds have been fired.  Typical erosion grooves produced by the propellant gases are present. X½.

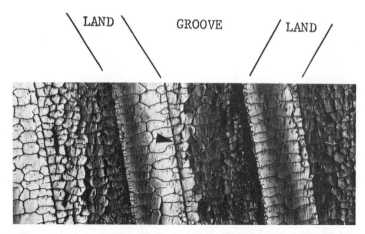

Fig. 3. Bore surface at shot-start region of a fired gun tube. Typical thermal cracks are present. Note also the continuous crack (arrowed) along the edge of the rifling X2.

Fig. 4. Longitudinal section of the bore surface of a fired gun tube. The lighter-etching layer at the surface is the thermally-affected layer. Two bore cracks are present. One (left) is a thermal crack confined to the thermally-altered layers. In the other (right) the thermal crack has extended into the tube wall by fatigue. Etched in picral. X100.

Nothing can at present be done metallurgically to reduce these effects, except perhaps to chromium plate the bore.  But this introduces other complications and will not be considered further.  It has been found, however, that bore erosion can be reduced considerably by using propellants with a low flame temperature or by adding certain chemicals to the propellant which deposit a thermally-insulating layer on the bore surface and so reduce the maximum temperature attained in the surface layers.  The end result is a considerable increase in the number of rounds that can be fired before an unacceptable degree of enlargement of the bore occurs.  This tends, however, to shift the life-limiting phenomenon into another area.  The thermal cracks still form and, particularly those located at the root corners of the rifling bands,  grow under the influence of the high hoop stresses induced in the gun tube when each round is fired (Figs. 4 and 5).  The tube is in effect subjected to high-stress low-cycle fatigue.

One of two things may then occur.  First, a crack may extend completely through the wall thickness of the tube which will cause the gun to malfunction but it will remain safe (leak before burst condition).  Secondly, a crack may grow to critical size and initiate catastrophic failure.  The second event, which must be avoided at all costs, has become increasingly likely with recent trends towards the use in gun tubes of steels of higher strength, and consequently of reduced fracture toughness, in order to reduce weight.

Consequently, fracture toughness and fatigue crack growth rate characteristics have become of great importance in modern gun steels.  The most effective way of

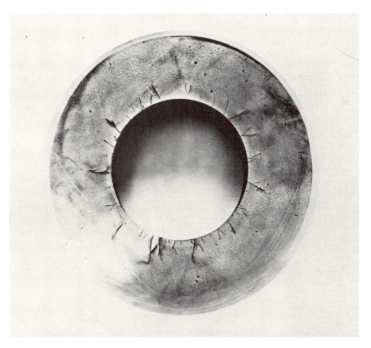

Fig. 5.  Section of a used gun tube, after magnetic particle inspection.  Fatigue cracks have grown approximately radially into the wall of the tube.  X⅓.

improving both parameters is to reduce the content of non-metallic inclusions, principally by reducing the sulphur content, and to improve the uniformity of distribution of the non-metallic inclusions that are present. Both objectives can be achieved by using the electroslag refining (ESR) process to produce the forging ingot. An example of the resultant reduction in fatigue crack growth rate compared to an air-melted steel of even exceptionally good quality is given in Fig. 6[2].

The base level content of non-metallic inclusions in these modern high-quality steels is so low that it is difficult to assess the inclusion content by metallographic methods, and hence to use them for quality control. However, even these steels may contain large exogenic inclusions arising from accidental contamination of the steel-making melt. Such inclusions are serious if exposed at the bore surface by machining (Fig. 7a) because severe local scouring may occur around such inclusions during firing. It is a straightforward exercise in optical metallography to confirm the presence of inclusions and to identify them (Fig. 7b). The presence of such inclusions would cause little concern in many heavy-duty forgings. The difficulty is, moreover, that an inclusion may be exposed only at a late stage of machining, when it can become cause for rejection of a tube after much time and money has been spent. It is normal practice, therefore, to inspect gun tube forgings by ultrasonic methods before machining is commenced.

The metallographic interest with these improved gun steels of low sulphur content centers in the phenomena of overheating and burning. *Overheating* is a term applied to any austenitizing treatment that impairs the properties of a steel. If normal properties cannot be restored by further heat treatment, the steel is then said to have been *burnt*. However, these terms are imprecise and at best generic covering a number of discrete and different phenomena.

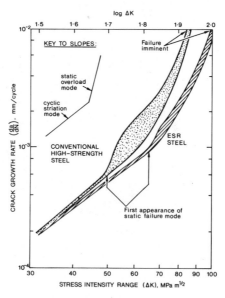

Fig. 6. Comparison of the rate of growth of fatigue cracks in comparable quenched-and-tempered gun steels, one made by conventional electric-arc air melting and one by the ESR process.

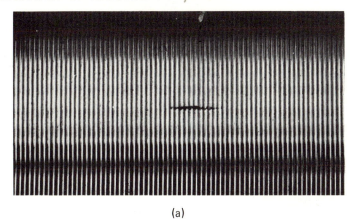

(a)

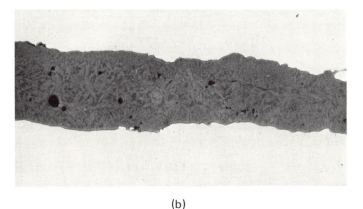

(b)

Fig. 7. Bore surface of a gun tube at a late machining stage. (a) A longitudinal marking on the surface. X2. (b) A metallographically-prepared surface of this region, indicating that the marking is an exogenic non-metallic inclusion. X100. This inclusion would have scoured out during firing.

The first phenomenon is *grain coarsening* — the production of an excessively large austenitic grain size. The second is *sulphide spheroidization* — partial or complete spheroidization of the normal manganese sulphide inclusions. The third is *grain-boundary sulphide precipitation* — the partial solution of the manganese sulphide inclusions at the austenitizing temperature followed by precipitation of particles of manganese sulphide at the austenitic grain boundaries during subsequent cooling from the austenitizing temperature. And, finally, the fourth is *grain-boundary liquation* — the melting of material in the regions of the austenitic grain boundaries followed by the solidification in situ of this material during cooling from the austenitizing temperature. These phenomena occur progressively in the order given as the forging temperature is increased.

The estimation of austenitic grain size by metallographic methods is a standard procedure. Spheroidization of sulphide inclusions can also readily be observed, an example being given in Fig. 8; it is not of itself a significant phenomenon but is a precursor to, and may occur simultaneously with, those now to be described.

Fig. 8.   An example of spheroidization of a sulphide inclusion.   A 0.5% carbon steel heated to 1350°C.   Etched in picral. X500.

Grain-boundary sulphide precipitation, on the other hand, can be detected only by the use of special metallographic procedures which can be applied after the steel has been re-austenitized, quenched and tempered.   Moreover, it is a phenomenon that is of special importance in steels of low sulphur content because the austenitizing temperature at which it may be induced is lower than for steels of conventional sulphur content.   It may then occur as much as 200°–300°C below the solidus temperature of the steel which will be only slightly above, or even in, the normal forging range.   Consequently, it can develop in these steels if control of forging temperature is only slightly unsatisfactory.

The phenomenon is likely first to be noted by the appearance of matte granular facets on fracture surfaces produced by notched impact which may cover the entire fracture surface or be mixed with areas of normal fibrous fracture.   An example of a mixed fracture is shown, as a scanning electron micrograph, in  Fig. 9a;  a typical granular facet is marked A and an area of fibrous fracture is marked B.   The intergranular (with respect to austenite) nature of the A regions and the transgranular nature of the B regions can be confirmed by examination of sections of the fracture surface (Fig. 9b).

However, examination at higher magnifications shows that both regions are covered by the dimples that are characteristic of ductile fracture, each dimple having been developed at a manganese sulphide inclusion (Figs.9c and 9d).   The only difference between the granular and fibrous areas is that the dimples are more numerous and their nucleating inclusions are smaller in the former case.   The conclusion is that the granular facets are areas of normal ductile fracture but the fracture path has been determined by arrays of small manganese sulphide inclusions on the austenitic

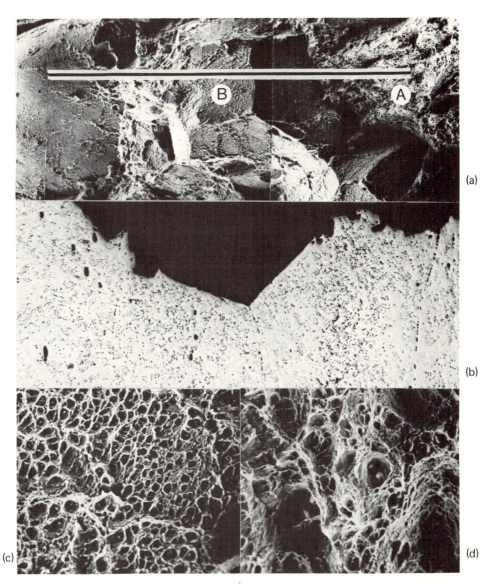

(a)

(b)

(c)

(d)

Fig. 9. An example of grain boundary sulphide precipitation. A high-strength gun steel has been heated to 1400°C, quenched and tempered. (a) Fracture surface. SEM. X100. (b) Section of the fracture surface along the line indicated in (a). Etched in nitric-sulphuric acid reagent. X100. (c) SEM of the fracture area marked A in (a) and (b). X500. (d) SEM of the fracture area marked B in (a) and (b). X500.

grain boundaries. They were precipitated there during cooling from the peak austenitizing temperature following solution of at least some of the primarily manganese
sulphide inclusions at peak temperature[3—6].

The sulphide inclusions responsible are too small to be detected directly in sections by optical microscopy, but their presence can be established indirectly by etching with specific reagents. The most reliable of these is a mixture of nitric and sulphuric acids,[7*] which develops chains of pits along the grain boundaries on which
the sulphide precipitation has occurred; the pits are best seen if the section surface
is repolished a little after etching to remove the general background etch. The specimen illustrated in Fig. 9b was treated in this way, and it is apparent that the pitting has developed along grain boundaries on which granular fracture would have occurred. The association between these chains of etch pits and the grain boundary
sulphides is indirect. Their direct cause has not yet been established; certainly,
individual sulphide particles are not delineated (cf. Figs. 9b and 9c). This is, however, unimportant for quality control.

In spite of its dramatic metallographic characteristics, grain boundary sulphide
precipitation does not cause any significant deterioration in mechanical properties,
such as quasi-static mechanical properties, fracture toughness, or fatigue crack growth
rate[8,9]. Grain-boundary liquation, on the other hand, does have significant effects
on properties. Consequently, it is of importance for quality control purposes to be
able to distinguish positively between the two, and this can be done by optical microscopy.

Grain-boundary liquation commences at a temperature below, but only slightly
below (~50°), the solidus temperature of a steel; that is, it occurs in a temperature
range between that at which grain-boundary sulphide precipitation occurs and the solidus. Manganese and sulphur segregate to the austenitic grain boundaries when a
steel is heated to a temperature at which primary manganese sulphide inclusions
dissolve. This reduces the melting point in the region of the grain boundaries to
below that of the nominal solidus temperature. Molten material, once it starts to
form, absorbs phosphorus from the surrounding austenite, because of the high parti-
tion coefficient of phosphorus in the system, further reducing the melting point in
the region. Flat primary dendrites of manganese sulphide form from the molten material when it solidifies, and the adjoining resolidified iron-rich regions in the region
of the grain boundaries are rich in phosphorus.

Thus typically, a diffuse differentially-etched band is observed along the austenitic grain boundaries of a liquated steel when it is subsequently austenitized,
quenched, and tempered (Figs. 10a and 10b), whereas this is never observed after
grain-boundary sulphide precipitation. Moreover, manganese-sulphide inclusions can
this time readily be detected along the grain boundaries by optical microscopy (Fig.
10b), whereas this is never so after grain boundary sulphide precipitation. The particles observed are sections of the arms of the flat dendrites of manganese sulphide
mentioned earlier. Fissures may also be present (Fig. 10c), although this is not
necessarily so.

Liquated steels fracture along the austenitic grain boundary to develop matte
granular facets similar to, but generally brighter than, those developed in steels in
which sulphide precipitation alone has occurred, and the facets are again found to
be dimpled when examined by scanning electron microscopy. The inclusions in each
dimple are larger than for sulphide precipitation, but the difference is not definitive.

* 10% $HNO_3$, 10% $H_2SO_4$ in distilled water. Etch by repeated immersion, swabbing off the
black surface deposits between immersions.

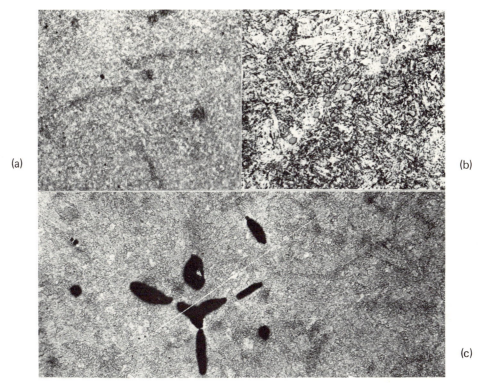

(a)

(b)

(c)

Fig. 10. An example of grain boundary liquation. A gun steel, the same as that illustrated in Fig. 9, has been heated to 1450°C, cooled to room temperature, reaustenitized, quenched and tempered. (a) Differentially etched regions at the high-temperature austenitic grain boundaries. Etched in picral. X50. (b) Sulphide particles at these austenitic grain boundaries. Etched in picral. X500. (c) Cavities at the austenitic grain boundaries. Etched in picral. X50.

Chains of pits may also be developed along liquated boundaries by etching in the nitric-sulphuric acid reagent; there are differences from the pits developed in material in which sulphide precipitation alone has occurred, but it would be difficult to distinguish between the two conditions on these grounds.

Summarizing, material in which either sulphide precipitation or liquation has occurred will:
(i)   have matte granular facets on fracture surfaces, particularly those produced by notched impact, and
(ii)  develop chains of etch pits along the austenitic grain boundaries when etched in a specific etching reagent.
But only in material in which grain boundary liquation has occurred will:
(i)   chains of manganese sulphide particles, and
(ii)  a differentially-etching band
be detected at the austenitic grain boundaries by optical microscopy.

The presence of grain-boundary liquation is good reason for rejecting a gun barrel forging, whereas grain-boundary sulphide precipitation is not, even if unusual and apparently frightening granular facets are apparent on fracture surfaces of test pieces taken from the forging.

## SHELL BODIES

There are two critical, if brief, phases in the operational life of a shell body. In the first, during launching, the body must maintain structural integrity whilst being subjected to the high accelerating forces because even partial collapse might initiate the catastrophic event of premature initiation of the explosive filling. Furthermore, the body must not contain any abnormalities that might themselves cause premature initiation of the explosive filling. The second critical phase is when the shell arrives at the target. The body may then be required to break up into fragments of a designated size, optimized to defeat an intended target. We shall discuss the launch phase only.

Shell bodies are designed to have strength adequate to resist direct firing stresses, and this can be assured by standard metallurgical practices. Our concern in the present discussion will consequently be confined to those types of abnormality that might adversely affect launch performance, concentrating on features that might initiate premature ignition of the explosive filling as being special to the topic.

Consider first forged shell bodies, which are made by hot piercing a bar followed by hot drawing. A peculiarity of this forging sequence is that material at the base of the shell body receives very little reduction (Fig. 11), and that the grain of the steel extends largely longitudinally through the base of the body. On the other hand, the material adjacent to the surface layers of the cavity is drawn extensively downwards from the nose region towards the base region.

A consequence of the first feature is that any central pipe-like discontinuities present in the original bar stock are not closed during forging, but may extend through the base of the final shell body (Fig. 12). There is then a risk that the propellant gases will erode through this weakness in the base and initiate the explosive filling within the shell, with disastrous consequences. Metallographic control to eliminate all pipe-containing bars from the feed stock is therefore essential in any shell forging shop. Although this is done assiduously in production, it must be admitted that shell designers do not yet trust metallurgists fully, and so attach a socalled *base plate* cut from rolled sheet or plate to the outer surface of the base of the shell (Fig. 1b) so to eliminate all possibility of end grain in this region. The attachment of a base plate introduces its own problems, but that is another story.

A consequence of the second feature of this particular forging process is that discontinuities in the bar stock are likely to be opened out at the surface of the cavity in the forging. The cavity has a complex shape (Fig. 1b) and machining after forging is avoided if possible. Consequently, depressions and the like introduced during forging may remain in the explosive-filled round. The concern is that any significant depressions left in the cavity surface may entrap pockets of air, these air pockets may be heated by adiabatic compression by set-back forces during firing, and initiate the explosive filling prematurely. This problem may well be more imaginary than real, because the cavity is usually lined with a varnish, but no one is prepared knowingly to take the risk.

There are any number of metallographic causes for the development of depressions and like irregularities in the cavity surface, and it will suffice to describe two of interest.

The first example concerns the effects of the hot-piercing and drawing sequence on non-metallic inclusions and like internal discontinuities. Large inclusions present

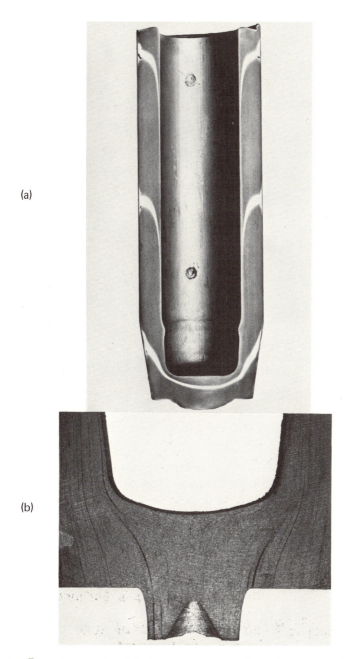

Fig. 11. Flow pattern developed during the piercing of a shell body. (a) Section of a pierced body; inserts set radially into the bar stock indicate the distribution of longitudinal flow. (b) Deeply-etched section of the base of a partly-machined forging, indicating the nature of grain flow in base region.

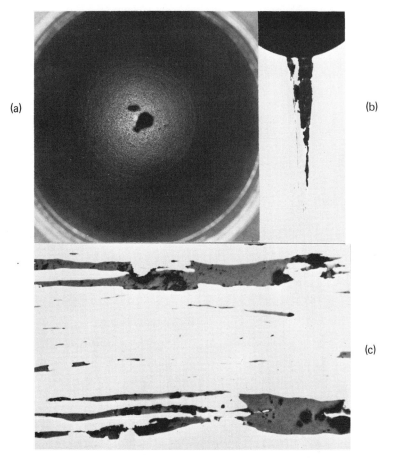

(a)

(b)

(c)

Fig. 12. Effect of a pipe in the bar stock used for a shell forging. (a) An open hole at the base of the cavity in a shell body forging. X1. (b) A longitudinal section through the hole illustrated in (a). X1. (c) Details of the non-metallic inclusions comprising the pipe. X100. Note that the pipe has opened out at the inner cavity surface, as is to be expected from the grain flow pattern shown in Fig. 11.

in the bar stock tend to fragment as they are drawn down the inner wall of the shell blank and open to produce seam-like defects when they were located close to the cavity surface, particularly towards the mouth (Fig. 13). Seams of this nature are quite unacceptable in this application. Certainly, the original inclusions in the example illustrated were larger than normally is to be expected in bar stock of this nature, but the example indicates that more stringent control of non-metallic inclusions than usual is necessary with feed stock for shell body forgings.

Other internal defects, such as hydrogen flakes, can have similar effects, and consequently precautions must be taken to ensure that they are not present in the forging bar stock. Moreover, the bar stock commonly is cut to length by nicking the bar and then fracturing it. Any laminations that develop in the fracture surface will produce seams in the cavity surface. These laminations may be associated with identifiable metallurgical abnormality, but they may not be. In a sense, a nicked-fracture test is carried out, and this can be used as a metallographic quality control test.

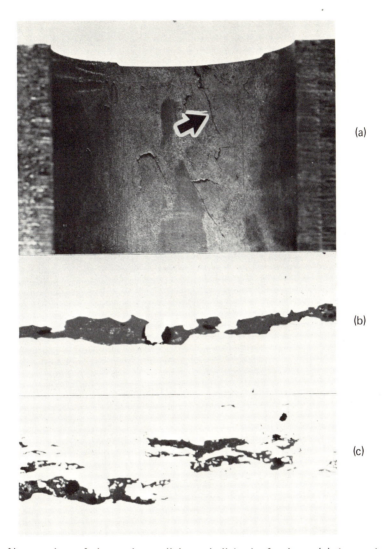

Fig. 13. Nose region of the cavity wall in a shell body forging. (a) Internal surface of the cavity, as forged. X1. (b) A stringer of non-metallic inclusions in the body of the forging. X500. (c) A fragmented stringer at the base of one of the seams shown in (a). X500. The inclusions have split during piercing to form seams in the cavity surface.

The second example arises because carbonaceous lubricants are used to facilitate the piercing operation. It is then found on occasions that globules of material with a carbon content as high as 4% may form at the base of the pierced cavity (Fig. 14). Regions with a carbon content greater than about 1.5% would be molten at the forging temperature and the globules of molten material so formed run down the side wall of the cavity and solidify when the pierced blank is inverted and cooled, as is normally done between piercing and drawing. The run of what is now a white cast iron is then pressed into the wall during the drawing stage. The run ruptures, so producing a series of fissures in the cavity wall (Fig. 14). Cavities of this nature are again quite unacceptable in a shell body. It appears that the processing variable that needs most to be controlled to prevent the formation of high-carbon regions is the preheat temperature. Small increases in temperature above the normal forging range greatly facilitate the rapid absorption of carbon into the steel.

## CARTRIDGE CASES

Much is required of the metal in a cartridge case. It must first of all be capable of being drawn into a tube with a large aspect ratio. It must not corrode or deteriorate during storage and must not react with the propellant filling. In use, it

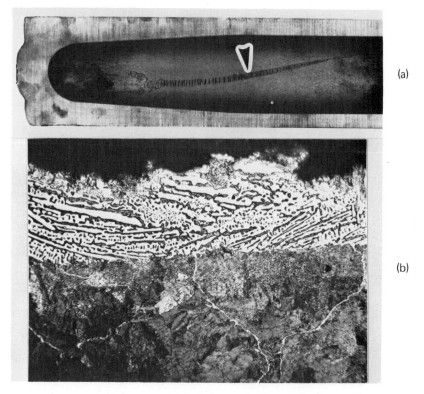

(a)

(b)

Fig. 14. Tears in the surface of the cavity in a shell body forging. (a) Inner surface of the cavity. X¼. (b) A transverse section through the band, indicated by the arrow, containing tears. Etched in picral. X100. Carbon from the piercing lubricant has been absorbed into the steel, to form brittle regions of a white cast iron.

must be capable of expanding to fill and seal the gun chamber when the propellant is ignited, and it must do this without splitting and without risk of reaction with the propellant gases. Moreover, it must remain sufficiently elastic to recover after the round has been fired to permit easy extraction from the gun chamber. The mouth of the case has to be soft so that it may be rolled to form a seal.

Only two alloys have so far proved to be capable of meeting these requirements. One is the classic cartridge brass (approx. 70%Cu, 30%Zn) and the other is steel, typically a medium-carbon grade in the quenched-and-tempered condition. The choice between these two alloys depends on the production base available (larger presses are required to draw steel cases), economics (copper alloys are intrinsically more costly than steels) and strategic availability. Quality control is usually based on specification of a hardness gradient required along the length of the case, and metallographic investigations can be invaluable in establishing what this gradient must be.

An example of the special problems encountered is given in Fig. 15, which is a photograph of the internal surface of a large-calibre brass cartridge case after firing. The base region of a number of cases of a prototype batch had separated during firing and, in many more such as that illustrated in Fig. 15a, a circumferential groove had developed in the inner surface at the position where base separation was occurring. Metallographic examination of sections through the groove (Fig. 15b) established that the groove and the associated deformation etch markings in the adjoining regions of the wall were characteristic of the necking-down prior to ductile tensile failure when this occurred under the simultaneous presence of hydrostatic pressure[10]. It could thereby be concluded that greater ductility was required in the wall material to withstand the strains imposed during firing. Cases annealed locally in this region to a lower hardness performed satisfactorily both in proof and subsequently in service. In general, the hardness gradient required in a cartridge case has to be set by specific experience in this way.

Brass cases that have to be softened at the mouth so that they may be rolled into the cannelure in a shell body are locally mouth annealed, and this often is effected by comparatively crude heating methods, such as by a gas flame. Under-annealing can be checked by hardness testing but over-annealing can be established only by metallographic examination. An example is given in Fig. 16, where partial liquation has occurred during mouth annealing, without visible external signs. These cases split at the mouth during firing.

Metallographic control to ensure the absence of inhomogeneities in the plate stock that might produce laminations in the drawn case is also necessary. For example, non-metallic inclusions may be found in hot-rolled brass plate. Moreover, due to inadequate control of composition and ingot solidification, regions containing beta phase may even be found. Inspection for these features can be made by means of simple notched-break tests, undesirable inhomogeneities resulting in the development of longitudinal lips in the fracture; an example is given in Fig. 17.

Steel cartridge cases may be drawn from slugs cut as transverse slices of the bar stock. Somewhat the same end-grain problems may then be encountered in the base as for the shell bodies discussed earlier. Although penetration of the base by the propellant gases is much less likely and in any event would not be as serious in this case, there is nevertheless a reluctance to accept for service cases which have discontinuities that appear to extend longitudinally through the base. An example of metallographic interest is given in Fig. 18, where again the abnormalities have arisen because of the presence of non-metallic inclusions in steel.

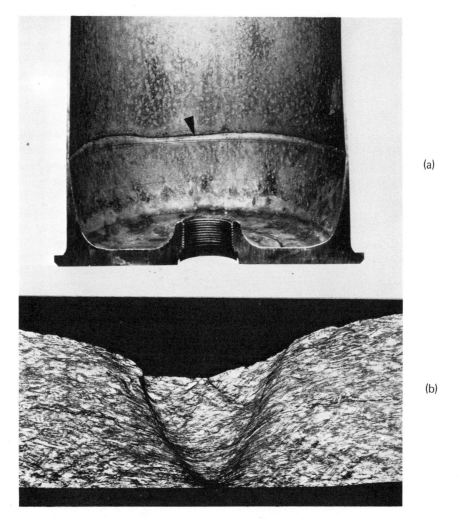

Fig. 15. Example of base separation during firing of a brass cartridge case. (a) Inner surface of the case, showing a circumferential groove developed during firing. X3/5. (b) Longitudinal section through this groove. Etched in ferric chloride reagent. The case wall has necked down assymetrically. Note the shear-band etch markings associated with the neck.

Steel cartridge cases are cadmium plated for corrosion protection during storage, which requires pre-treatment by acid pickling. Pits may develop in the surface of the base during the pickling treatment, particularly if it has to be repeated. These pits will be visible in the finished part (Fig. 18a), particularly after a period of storage when remnants of the electroplating solutions tend to seep out of the pits and cause local corrosion. The pits are initiated at, and develop along, manganese sulphide inclusions in the steel (Fig. 18b). The interesting thing is that pitting of this nature is nucleated by preferential attack at the *junction* between the inclusion and the steel matrix (Fig. 19).[11]   Presumably there are discontinuities between the inclusions and the matrix along which the etching solutions can penetrate.

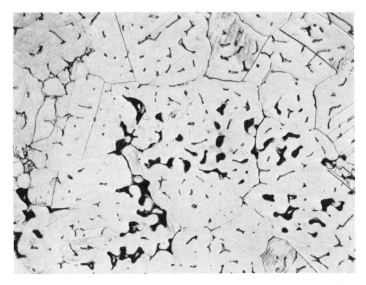

Fig. 16. Microstructure of the mouth end of a brass cartridge case. Partial liquation has occurred during mouth annealing. Etched in a ferric chloride reagent. X500.

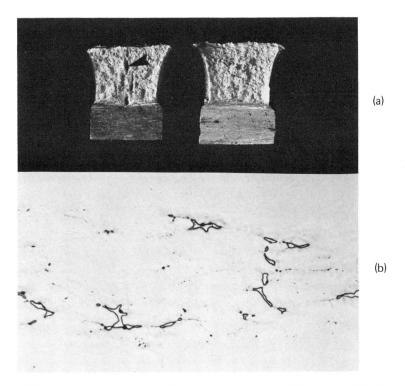

(a)

(b)

Fig. 17. (a) Nicked fractures in plate blank for a brass cartridge case. (b) Microstructure of material at left. Some beta phase is present and is responsible for the lips developed in the fracture surface. Etched in a ferric chloride reagent. X500.

(a)

(b)

Fig. 18.   Etching pits developed in the base of a cold-drawn steel cartridge case.
(a) Appearance of the pits.   X2.   (b) Longitudinal section of a pit.   Etched in nital.
X500.   The pits have developed along a stringer of manganese sulphide inclusions
(arrowed), but do not extend far into the base.

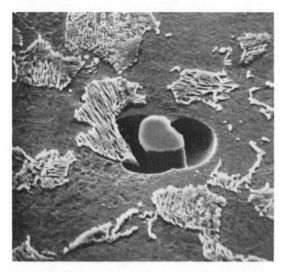

Fig. 19.   The base surface of the cartridge case illustrated in Fig. 17.   The surface
has been metallographically polished and then etched in an iodine/potassium iodide
solute.   Scanning electron micrograph.   X500.   Note that the etching attack has occurred along the interface between the manganese sulphuric inclusion (centre) and
the ferrite of the matrix.

REMARKS

Many of the examples illustrate that stricter attention than usual has to be given to the quality control of the metals used in armament hardware because of the possibility that material defects may have catastrophic consequences during the use of the equipment. The probability of these events actually occurring may be very small but risks of even this magnitude are not taken knowingly, not only because of the consequence of the event itself but because the morale of military personnel is related to the confidence that they have in the equipments that they have to use. The metallographer plays a key role in both establishing and maintaining the required quality standards.

ACKNOWLEDGEMENTS

This paper is based on investigations carried out so competently by many members of the staff of Materials Research Laboratories, too numerous to name, over a number of years.

REFERENCES

1.  R.B. Griffin, J. Pope and C. Morris, *Metallography,* 8, 453 (1975).
2.  J.C. Ritter and M.E. de Morton, *J. Aust. Inst. Metals,* 22, 51 (1977).
3.  R.D. Haworth, Jr. and A.F. Christian, *Proc.* ASTM, 45, 407 (1945).
4.  H.J. Merchant, *J. Iron Steel Inst.,* 45, 217 (1946).
5.  G.D. Joy and J. Nutting, *Proc. ISI/BISRA Conf.,* "Effects of Second Phase Particles on the Mechanical Properties of Steel". The Iron and Steel Inst., Scarborough, 95 (1971).
6.  R.C. Andrew, G.M. Weston and R.T. Southlin, *J. Aust. Inst. Metals, (1976).*
7.  G.W. Austin, *Iron Steel Inst.,* Special Report No. 14, 189 (1936).
8.  R.C. Andrew and G.M. Weston, *J. Aust. Inst. Metals,* 22, 171 (1977).
9.  R.C. Andrew and G.M. Weston, *J. Aust. Inst. Metals,* 22, 200 (1977).
10. I.E. French and P.F. Weinrich, *J. Aust. Inst. Metals,* 22, 40 (1977).
11. V.M. Silva and L.F.G. Williams, *Surface Technology,* 6, 131 (1977).

# METALLOGRAPHIC QUALITY CONTROL
## OF ORTHOPAEDIC IMPLANTS

G. Hamman and D.I. Bardos*

## INTRODUCTION

Metallography plays a significant role in the quality control of metal alloys used in the manufacture of orthopaedic surgical implants. Metallographic examination of raw materials prior to the fabrication of implant devices helps assure that the devices will be safe and effective when used in the treatment of orthopaedic patients. In most instances, the devices must have adequate mechanical strength, fatigue strength, corrosion resistance, and biocompatibility. While it is not possible to predict the clinical performance of materials by microstructural analysis alone, metallography does serve to relate the clinically-observed performances of materials to the metallurgical characteristics of the raw materials. This paper highlights the important aspects of this relationship as observed during many years of examining thousands of devices.

## RAW MATERIAL ACCEPTANCE

Raw materials reaching implant manufacturers have already undergone numerous quality control examinations during the alloy formulation and mill product fabrication stages. Metallurgical acceptance of materials varies with particular alloys. In the case of cold-worked stainless steel for the highest-quality implant devices, the metallurgical and chemical properties of the raw material are not altered during the device fabrication stages. Therefore, the metallographic requirements for the raw material are the same as those required for the final product. Similarly, Ti-6Al-4V titanium alloy in the wrought, mill-annealed condition is used to fabricate implants. Forgings prepared from these two alloys receive their final metallographic quality control inspections subsequent to the forging process. However, with Co-Cr-Mo cobalt casting alloy, metallography is not used alone as raw material acceptance criteria since this material is subsequently remelted and investment cast into final product shapes. Co-Cr-Mo parts can also be fabricated by powder metallurgy technology, or hot working of the cobalt-base alloy, and these formed pieces are examined metallographically and acceptance criteria are established for them also.

## STAINLESS STEELS

American Society for Testing and Materials (ASTM) specifications provide acceptance criteria for such stainless steel implant material microscopic features as: grain size, nonmetallic inclusions, and microconstituents such as ferrites and carbides.

*Zimmer, USA, Warsaw, Indiana 46580 USA.

Grain size determinations are performed in accordance with ASTM Standard E-112, Recommended Practices. The etching methods for all polished stainless steel specimens examined in the authors' laboratory are as follows:

Electrolyte — 10% aqueous solution of ammonium persulfate
Voltage — 4 VDC
Time — 45 seconds
Surface Area — approximately ½ square inch

For cold-worked stainless steel, it is recommended that transverse sections be prepared to avoid possible grain size determination errors arising from the slightly elongated shape of the cold-worked equiaxed grains. Figure 1 illustrates a transverse section of moderately cold-worked stainless steel. The grain size was determined to be No. 5.5. Figure 2 illustrates more highly cold-worked stainless steel and in the transverse section shown, the grain size was determined by the intercept method to be No. 6.25. In Fig. 3, the longitudinal section of the same sample, the intercept method yielded a value of No. 5.75. This microstructure corresponds to material with an approximate ultimate tensile strength of 150,000 psi. In more severely cold-worked material, the difference between transverse and longitudinal sections may be more pronounced. In cold-worked stainless steel, grain size of No. 5 or finer is preferred for high-quality implant applications. Therefore, in most applications, a grain size of No. 4 or coarser is a rejectable condition.

Hot forging stainless steel for implant applications produces more varied grain structure. Figure 4 illustrates a uniform grain size of No. 5 resulting from hot forging and subsequent annealing at 1850°F. Figure 5 illustrates the cross-section of a forging that exhibits a rarely seen duplex grain size. Higher magnification of the structure is shown in Fig. 6. The duplex grain size may possibly be eliminated by higher temperature annealing, but this treatment may result in excessively-coarse grain size and is not recommended, therefore, as a corrective measure. The duplex grain structure has not been shown to have detrimental effects on the properties of the material which affect its clinical performance in orthopaedic devices. ASTM standard specification F621 established guidelines to provide for more uniform grain size and states that the grain size shal be ASTM No. 5 or finer, except that up to 10 percent of the cross-sectional area may have ASTM No. 3 or finer, provided there is no banding of coarse grain.

Similar duplex grain size has been observed in hot-finished, or moderately cold-worked bar stock. The duplex grain structure illustrated in Figs. 7 and 8 has been observed in bar stock over one-inch diameter. The classical picture of as-recrystallized grains adjacent to larger, cold-worked grains is clearly visible. Such larger cold-worked areas in a recrystallized matrix are not known to cause detrimental changes in the properties of the alloy and their presence does not, therefore, constitute a rejectable condition. If a larger region of the cross-section exhibits the very-coarse, cold-worked grains in contrast to other regions of the alloy with acceptable grain structure, then the material is less desirable for implant applications. As previously noted, these duplex grain structures are rare occurrences in the steel used in the fabrication of devices for the internal fixation of fractures and having a typical grain structure of ASTM No. 5.5, as shown in Fig. 1.

Occasionally material with macrostructure illustrated in Fig. 9 is encountered. The macrostructure shown indicates alloy segregation. At low-power magnifications, or by visual examination, the etched specimen exhibits obvious bands of lighter and darker regions. On the other hand, at 100X magnification, the demarcation between the lighter and darker bands is not as easily noted. In Fig. 10, the boundary between the light band and the darker band is depicted. The grain structure is uniform and

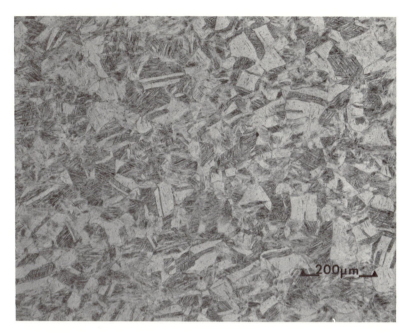

Fig. 1. Wrought 316L stainless steel (100X), transverse cross-section. Etchant: 10% ammonium persulfate, electrolytic.

Fig. 2. Cold worked 316L stainless steel (100X), transverse cross-section. Etchant: 10% ammonium persulfate, electrolytic.

Fig. 3.  Cold worked 316L stainless steel (100X), longitudinal section.  Etchant: 10% ammonium persulfate, electrolytic.

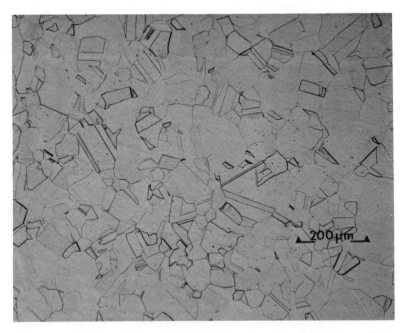

Fig. 4.  Hot-forged 316L stainless steel (100X), transverse cross-section.  Etchant: 10% ammonium persulfate, electrolytic.

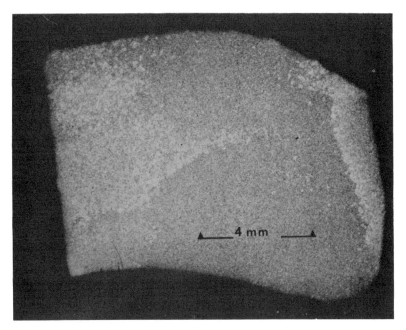

Fig. 5. Hot-forged 316L stainless steel (8X), transverse cross-section. Etchant: 10% ammonium persulfate, electrolytic.

Fig. 6. Hot-forged 316L stainless steel (100X), transverse cross-section (same specimen as Figure 5). Etchant: 10% ammonium persulfate, electrolytic.

Fig. 7. Large diameter wrought 316L stainless steel bar (8X), transverse cross-section. Etchant: 10% ammonium persulfate, electrolytic. (Reduced 25% for reproduction.)

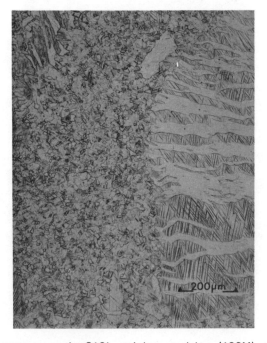

Fig. 8. Large diameter wrought 316L stainless steel bar (100X), transverse cross-section (same specimen as in Figure 7). Etchant: 10% ammonium persulfate, electrolytic. (Reduced 25% for reproduction.)

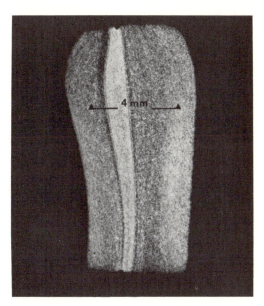

Fig. 9. Hot-forged 316L stainless steel (8X), transverse cross-section. Etchant: 10% ammonium persulfate, electrolytic.

Fig. 10. Hot-forged 316L stainless steel (100X), transverse cross-section (same specimen as in Figure 9). Etchant: 10% ammonium persulfate, electrolytic.

the grain size is unaffected by the mild alloy segregation. The etchant preferentially attacks regions where the composition differs slightly from the matrix. This attack manifests itself in the dark streaks in the rolling direction. These streaks are not to be mistaken for non-metallic inclusions. The electrolytic etching has a far more severe attack on the surface of the metal than encountered in service. Therefore, it is not expected that the mild alloy segregation depicted in the above figures is likely to have a detrimental effect on the performance, i.e., the corrosion resistance of the alloy. Samples of such material with similar microstructural features, removed from human body environments, have not shown signs of corrosive attack.

The classic problem of sensitization with stainless steel of higher carbon content is rarely encountered. Figure 11 depicts a rare sample of fully-sensitized, AISI 316L stainless steel. The 1000X magnification of Fig. 12 shows the mildly-cold-worked grains with slip bands and string-of-pearl appearance of the carbide particles outlining the grain boundaries. Such a fully-sensitized condition is clearly not acceptable due to the susceptibility of this type material to corrosion.

In samples of forgings, scattered clusters of carbides are occasionally encountered. Figure 13 shows such a region with heavy, carbide-particle precipitates. Since these carbides do not form a continuous network, they do not have the same detrimental effects as the fully-sensitized material. Therefore, this condition is acceptable under existing ASTM material specifications.

Ferrites in the microstructure of 316L stainless steel are considered to be associated with reduced corrosion resistance in the environment of the human body. ASTM material specifications clearly state that during the routine examination of raw material at 100X, no ferrites shall be detected. In examining samples during the 1970 decade, no ferrites have been observed by the authors in 316LVM stainless steel accepted for surgical implant manufacture. A rare sample of a prosthesis of unknown manufacture shown in Fig. 14, exhibited extensive ferrite content. This sample was prepared under the usual procedures and etched with Murakami's reagent with the following procedure. The polished specimen was immersed for 90 seconds in a boiling solution of 30 g $K_3Fe(CN)_6$, 30 g KOH and 60 ml $H_2O$, then water rinsed, ultrasonically cleaned, rinsed in alcohol and finally dried. The advantage of this etchant is that it stains ferrite a yellowish hue, while sigma phase is tinted bluish. This reagent provides easy identification of these two undesirable microconstituents.

Microcleanliness specifications for implant applications are outlined in ASTM Standards F55 and F138. "Grade 2" of F138 specification is the most widely used stainless steel implant material today. It permits a maximum allowable limit of non-metallic inclusions of 1.0 for the heavy series and 1.5 maximum for the thin series. Plate III of ASTM Standard E45 is usually used for these determinations. The most commonly used method for microcleanliness determinations utilizes "Method A" which depends on a comparison of worst field with the established standard. Such a combination of Method A with Plate III provides for rapid determination of nonmetallic inclusion level. In the vacuum-melted, low-carbon stainless steels, the most frequently enountered nonmetallic inclusions are sulfides. Figure 15 shows a typical field where minute quantities of sulfides are visible. Sulfide stringers are shown at high magnification in Fig. 16. In extreme cases, the sulfide content exceeds the permissible maximum levels. Such a rejectable case is illustrated in Fig. 17. In addition to sulfides, oxide particles are occasionally enountered. The inclusion content of today's steels is usually far below the permitted maximum levels. The effect of nonmetallic inclusions in such minute quantities on implant performance has not been established. It is believed that with high levels of nonmetallic inclusions, the fatigue

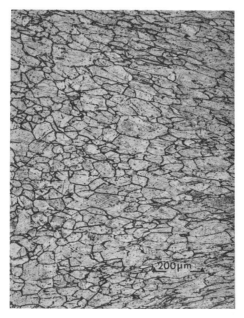

Fig. 11. Sensitized 316L stainless steel (100X), transverse cross-section. Etchant: 10% ammonium persulfate, electrolytic. (Reduced 30% for reproduction.)

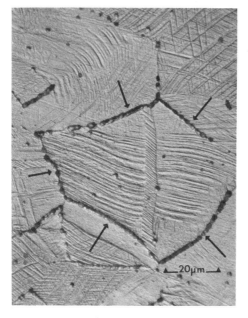

Fig. 12. Total ditch structure from Figure 11 (1000X). Etchant: 10% ammonium persulfate, electrolytic. (Reduced 30% for reproduction.)

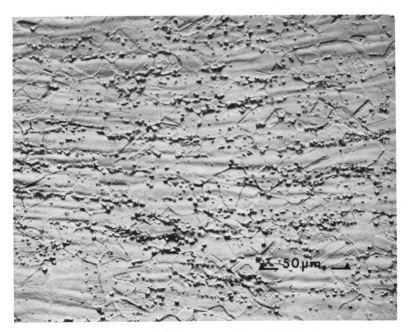

Fig. 13.  Precipitated carbides in 316L stainless steel (500X), transverse cross-section.
Etchant:  10% ammonium persulfate, electrolytic.

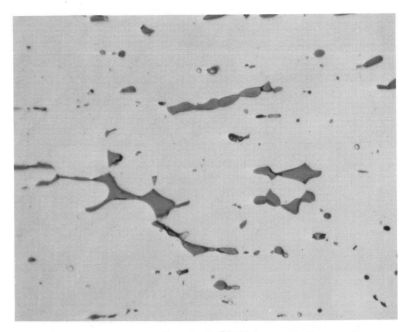

Fig. 14.  Ferrites in stainless steel (nominally 316L), sample from unknown manufacturer, (1000X), transverse cross-section.  Etchant:  Murakami's reagent.

Fig. 15.  316L stainless steel with low nonmetallic content (100X), longitudinal section.  As-polished.

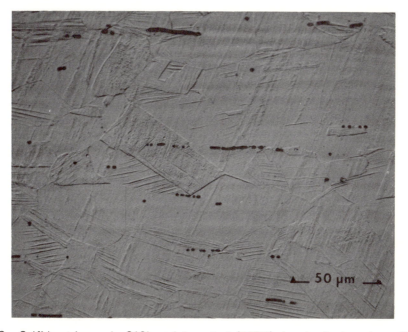

Fig. 16.  Sulfide stringers in 316L stainless steel (500X), longitudinal section.  Etchant:  10% ammonium persulfate, electrolytic.

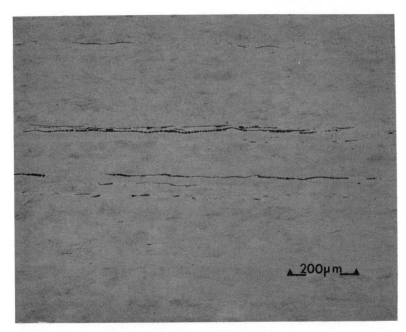

Fig. 17. Excessive nonmetallic content in 316L stainless steel (100X), longitudinal section. Etchant:  10% ammonium persulfate, electrolytic.

strength is decreased and the material may be more susceptible to corrosive attack by the extracellular chloride environment encountered in the human body.  Such a condition is illustrated in Fig. 18.  This figure shows a compression hip screw removed from a patient after two-years implantation.  In Fig. 19, numerous pits are evident under low-power magnification.  The surfaces of the threads were electropolished and passivated prior to implantation.  Suspected pitting corrosion was confirmed on metallographic examination of the device.  Figure 20 shows the cross-section of a thread with a corrosion pit marked by an arrow.  It is evident in this figure that pitting took place at the intersection of large non-metallic inclusions where they intersect the polished surface.  Such a minute quantity of corrosion would be well tolerated by the patient and would not usually have clinical significance.  In some patients, however, the corrosion products may provoke tissue inflammation in the area of the implant and the surgeon may elect to remove the device to alleviate the clinical symptoms.

## TITANIUM AND Ti—6Al—4V ALLOY

Titanium and its alloys provide a variety of microstructural features.  The quality control of titanium alloys by microstructure is very complex.  The pure titanium microstructure shown in Fig. 21 is typical of the fine-grain material used in some applications.  More commonly, ASTM F136 titanium alloy is used in implant applications.  In the usual wrought and mill-annealed condition, the microstructure is typically very fine grained with a duplex structure of small alpha grains as shown, appearing as white, in Fig. 22 with the beta phase appearing as the dark microconstituent.  This uniform, fine-grained structure is most desirable.  Quality control criteria are very difficult to establish for this alloy and frequently a careful evaluation of various microstructures are required, in combination with metallurgical and mechanical evalua-

Fig. 18. Close-up of threaded end of retrieved compression lag screw.

Fig. 19. Close-up of pitting corrosion on threaded area of compression lag screw.

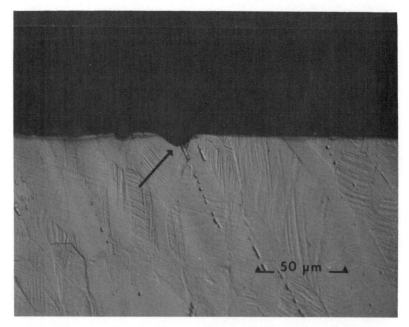

Fig. 20. Longitudinal section showing profile of pitted threads from Figs. 18 and 19. (500X). Etchant: 10% ammonium persulfate, electrolytic.

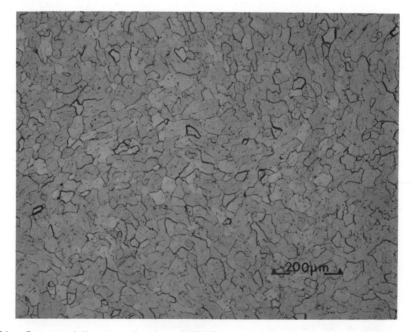

Fig. 21. Commercially pure titanium (100X), transverse cross-section. Etchant: 10% HF, 5% $HNO_3$. Immersion 10 seconds.

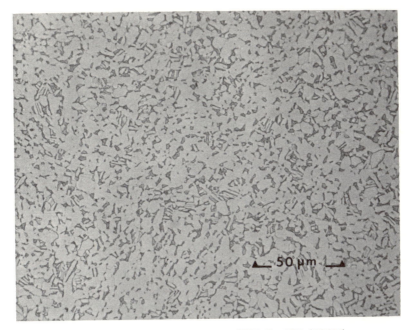

Fig. 22.  Mill-annealed Ti- 6Al—4V ELI, ASTM F—136 (500X), transverse cross-section.  Etchant:  10% HF, 5% $HNO_3$.  Immersion 2 to 4 seconds.

tion of the material, prior to acceptance.  In the hot-forged and annealed condition, the microstructure is still the alpha-beta structure illustrated in Fig. 23.  Improper thermal treatment results in an excessive, coarse alpha-platelet structure.  This condition is illustrated in Fig. 24.  An alpha case is unavoidable on the surfaces of forgings.  The forging shop usually removes the alpha case by mechanical or chemical means.

Figure 25 depicts the same alloy in the cast and annealed condition.  Large, prior, beta grain boundaries are evident, with the transformation structure in the basket-weave fashion.  In the cast condition, a variety of microstructures may be encountered.  A very-coarse, alpha network is thought to reduce the mechanical strength of the alloy.

## CAST COBALT—CHROME—MOLYBDENUM ALLOY

Cast cobalt-chrome-molybdenum alloy is commonly the material of choice for prosthetic applications.  The typical as-cast microstructure is frequently encountered. Figure 26 shows the uniform distribution of primary carbides in a very-large-grained cobalt base matrix.  The occurrence of the carbides and their stoichiometric and morphology conditions depend on the prior thermal treatment of the alloy.  In the heat-treated condition, precipitation of secondary carbides are occasionally observed. Metallography is not used as a quality control technique with cast cobalt-chrome alloy.  The microstructure does not lend itself to a ready association between performance and microconstituents.

Casting porosity is an area of concern with this alloy.  Nondestructive test methods, such as radiography and penetrant inspection, are commonly used to

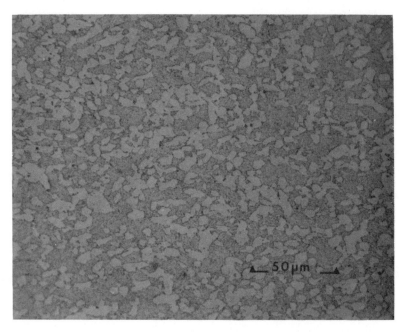

Fig. 23. Hot-forged, annealed Ti- 6Al—4V, ASTM F—136 (500X), transverse cross-section. Etchant:   10% HF, 5% HNO$_3$.   Immersion 2 to 4 seconds.

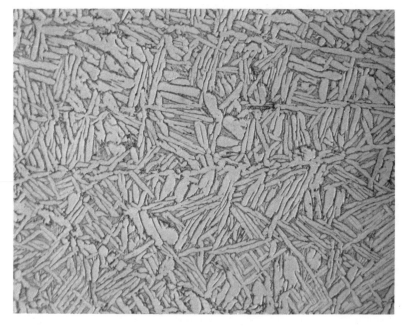

Fig. 24.  Coarse, elongated alpha platelets in hot-forged Ti—6Al—4V alloy (500X). Etchant:   10% HF, 5% HNO$_3$.   Immersion 2 to 4 seconds.

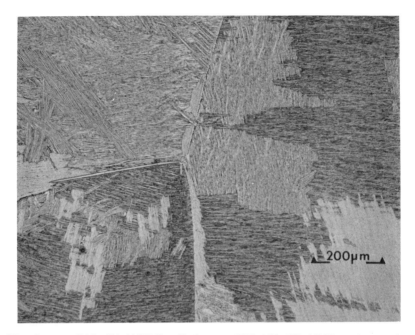

Fig. 25. Cast Ti-6Al–4V (100X). Etchant: 10% HF, 5% HNO$_3$. Immersion 2 to 4 seconds.

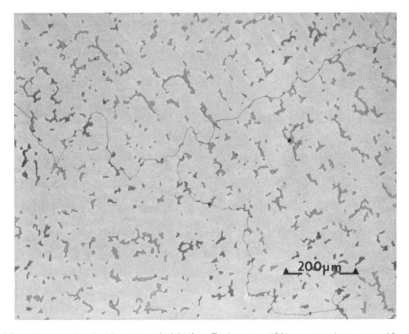

Fig. 26. As-cast Co-Cr-Mo alloy (100X). Etchant: 10% ammonium persulfate, electrolytic. 2 to 5 seconds, 4 volts DC, 1 amp/square centimeter.

detect casting defects.  Microporosity, on the other hand, is not detectable by NDT techniques.  Figure 27 shows a typical polished section of a cast cobalt-alloy prosthesis which is free of microstructural defects.  In Fig. 28, minute quantities of microporosity are evident.  Higher magnification of the condition shown in Fig. 29 reveals the typical interdendritic shrinkage porosity.

X-ray indications are occasionally confirmed by metallographic techniques. Figure 30 shows an X-ray image of a hip prosthesis.  The arrow marks the location of an indication which was the cause for the rejection of this part.  A metallographic study was performed to relate the X-ray image to microstructural features.  Figure 31 shows a cross-section of the prosthesis in the general region of the X-ray indication.  A section was made adjoining the approximate location of the indication. Grinding on a 120 grit belt was continued until the surface intersected the defect. The casting defect is shown by the arrow.  The high magnification picture in Fig. 32 shows the magnitude of the defect.  Such a severe defect is detectable by radiography and is, therefore, clearly a rejectable condition.

It is difficult to rate the cobalt alloy for nonmetallic inclusion content.  No standard rating exists for this material.  Figure 33 illustrates the worst case of foreign material inclusion found in a casting.  Radiography detected the presence of this inclusion and the casting was, therefore, rejected.

Fig. 27.  As-cast Co-Cr-Mo alloy (100X).  As-polished.

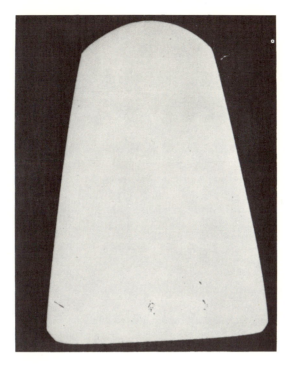

Fig. 28.  Microporosity in cast Co-Cr-Mo alloy (8x).  As-polished.

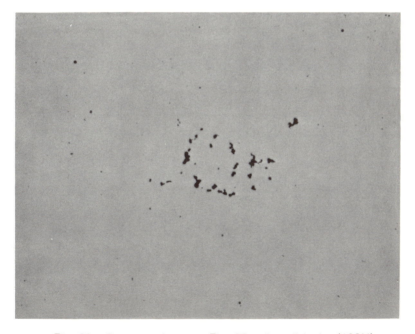

Fig. 29.  Same specimen as Fig. 28.  As-polished.  (100X).

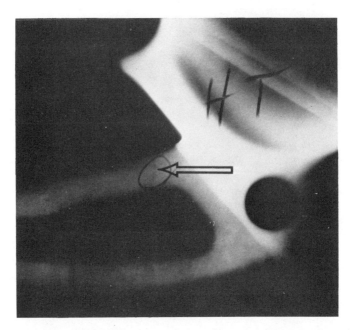

Fig. 30.  X-ray of cast Co-Cr-Mo hip prosthesis.  (Arrow marks location of X-ray indication).

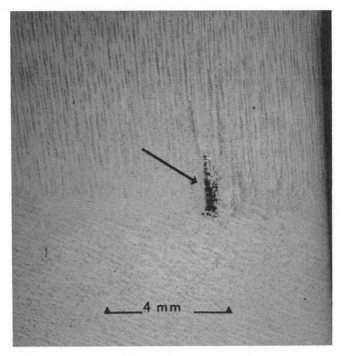

Fig. 31.  Metallographic cross-section through X-ray indication in Fig. 30 (8x).  As-polished.

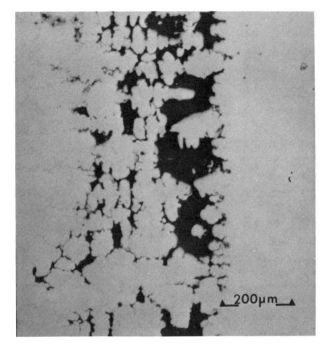

Fig. 32. Same specimen as in Fig. 31 (100X). As-polished.

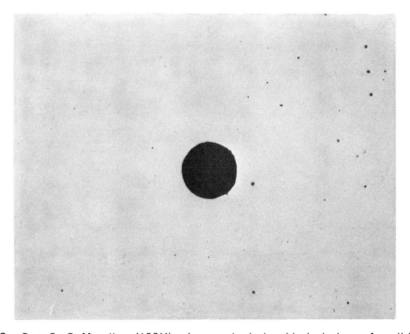

Fig. 33. Cast Co-Cr-Mo alloy (100X). Large spherical oxide inclusion. As-polished.

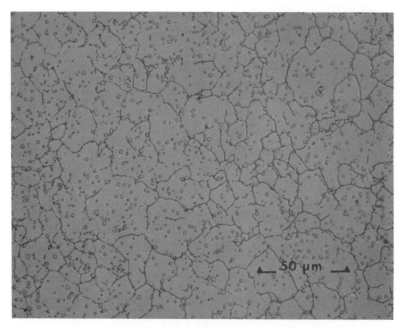

Fig. 34. Hot isostatically pressed Co-Cr-Mo alloy (500X). Etchant:   10% ammonium persulfate, electrolytic.   2 to 5 seconds, 4 volts DC, 1 amp/square centimeter.

Fig. 35. Polyethylene/graphite fiber composite. (100X). As-polished.

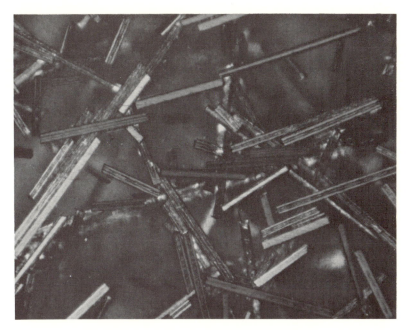

Fig. 36.  Same specimen as Fig. 35 (200X).

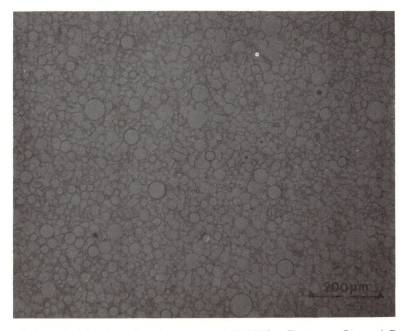

Fig. 37.  Polymethyl Methacrylate bone cement (100X).   Etchant:  Plasma/ Fuming $HNO_3$.

## ULTRA—FINE GRAIN COBALT—CHROME—ALLOY

Recent metallurgical innovations permit the fabrication of ultra-fine-grained cobalt-chromium-molybdenum alloy. A combination of powder metallurgy with hot isostatic pressing for the consolidation of the powder results in the extremely fine-grained structure as shown in Fig. 34. The uniform distribution of minute primary carbide particles in a cobalt base ultra-fine-grained matrix is especially noteworthy. Routine examination of such a fine-grained structure requires 500X magnification. The material shown in Fig. 34 has significantly improved mechanical properties and fatigue endurance limit. Metallurgical innovations to significantly improve the performance of orthopaedic devices was thus accomplished.

## OTHER ORTHOPAEDIC IMPLANT MATERIALS

Other materials used in orthopaedic applications do not normally lend themselves to metallographic quality control procedures. Metallography plays a significant role, however, in the research and development of new materials. The introduction of carbon-fiber-reinforced polyethylene as a bearing material in total joint applications provides significantly improved material properties for reduced wear and cold flow. Metallography played a significant role in this research effort. Figures 35 and 36 illustrate randomly distributed carbon fiber particles in a polyethylene matrix. For the fibers to effectively reinforce the polyethylene, the aspect ratio of the fibers must be maintained. Metallographic techniques detected breakage of the fibers from the nominal 1/8-inch lengths. Improved manufacturing techniques eliminated this problem, thus maintaining the desired length of the carbon fibers. Since the metallography of this material is not well known, it is not used as a quality control technique.

Bone cement has been successfully used for the fixation of total joint prostheses in the medullary canals of patients' bones. Polymethyl methacrylate self-curing material is used. The microstructure of typical polymerized bone cement is shown in Fig. 37. The material is supplied as polymer beads and liquid monomer and the two are mixed during the operation. The monomer dissolves the smaller particles and the surfaces of the polymer beads and polymerizes the entire mass within a few minutes. The figure shows the microstructure with the spherical, original polymer beads in a matrix of indistinguishable polymer with the same composition.

## REFERENCES

ASTM E3-62 (Reapproved 1974), "Standard Methods of Preparation of Metallographic Specimens".

ASTM E45-76, "Standard Recommended Practice for Determining Inclusion Content of Steel".

ASTM E112-74,"Standard Methods for Estimating the Average Grain Size of Metals".

ANSI/ASTM F75-76, "Standard Specification for Cast Cobalt-Chromium-Molybdenum Alloy for Surgical Implant Applications".

ANSI/ASTM F90-76, "Standard Specification for Wrought Cobalt-Chromium-Tungsten-Nickel Alloy for Surgical Implant Applications".

ANSI/ASTM F67-77, "Standard Specification for Unalloyed Titanium for Surgical Implant Applications".

ASTM F136-79, "Standard Specification for Titanium 6Al-4V ELI Alloy for Surgical Implant Applications".

ASTM F138-76, "Standard Specification for Stainless Steel Bar and Wire for Surgical Implants (Special Quality)".

ASTM F620-79, "Standard Specification for Titanium 6Al–4V ELI Alloy Forgings for Surgical Implants".

ASTM F621-79, "Standard Specification for Stainless Steel Forgings for Surgical Implants".

*ASM Metals Handbook,* Vol. 7, 8th Edition, "Atlas of Microstructures of Industrial Alloys".

D.I. Bardos and H.A. Luckey, "Microstructural Analyses of Biomaterials", Zimmer. USA, Inc. *Microstructural Science,* Vol. 3, (1975).

D.I. Bardos and M.E. Parker, "Microporosity in Cast Co-Cr-Mo Hip Prostheses", Zimmer.USA, Inc. *Microstructural Science,* Vol. 4, (1976).

# METALLOGRAPHIC EVALUATION OF MULTI—MATERIAL PRODUCTS

J.E. Dresty*, R.J. Orlowski and J.W. Lane

## INTRODUCTION

Multi-material products have become more commonly used in many high technology industries. With these products, properties can be obtained which neither the component materials individually nor the alloyed combinations can achieve. A composite material is a classic example; the final product has the strength of the fibers plus the toughness of the matrix. Another example of a multi-material product which combines the best of its material components is a plated part where the substrate material possesses good mechanical properties but lacks corrosion resistance, and the plating material provides the required corrosion protection. Metallographic examination of the interface between the fiber and the matrix, or the substrate and the plating, will reveal much information about the quality of the product.

Multi-material products can be manufactured by a number of processes of varying sophistication, including casting, powder metallurgy, hot isostatic pressing, roll bonding, welding and plating. For the purposes of this paper, only products with metallurgical bonds will be considered as multi-material; mechanically attached parts will not be considered. Evaluation of welds also will not be considered since weld metallography constitutes a huge body of knowledge suitable for a separate treatment. What will be discussed are some of the more unusual metallographic evaluations of multi-material products and some of the special problems associated with such evaluations.

## SAMPLING AND SECTIONING

The initial steps in metallographic evaluation of multi-material products are the definition of a sampling plan and the sectioning of the components. Standard procedures for metallographic sampling plans may be used such that sampling takes place at decreasing frequency as statistical confidence is gained. For repetitive metallographic evaluations, section lines may be scribed on the surface of the components using standard templates on areas coated with blue dye. The actual sectioning may be done with an abrasive cutter, EDM-abrasive cutter, metal band saw, or in some cases, even a shear. In sectioning of multi-material products there are special concerns not applicable to single material products. Strategic areas must be exposed by precise section location to properly display the internal features to be examined. Special care must be taken not to damage the features. For example, using a shear for sectioning may introduce excess work or mechanical damage to a feature which is sensitive to those actions, with the resultant distortion of date (e.g. fiber damage in a composite).

* Suisman and Blumenthal, Inc., Hartford, Connecticut 06101 USA.

Also, heat generated during abrasive cutter sectioning when excessive pressure is applied can modify the results of an evaluation of a heat-sensitive feature (e.g. grain structure change).

## PREPARATION

Metallographic preparation of multi-material products often presents special problems because of the need to polish materials of a very dissimilar mechanical properties in intimate contact with one another. Component parts of multi-material products are often either very hard or very soft and the interface between the two is normally a feature from which much information can be gathered concerning the quality of the product. Good edge retention of the interface is necessary to obtain accurate data. Typically, preparation techniques which provide high pressure and a hard polishing surface are necessary for providing flatness in multi-material products. Lapping techniques and automets with glass surfaces and canvas cloths have been found to be the most effective.

Preliminary polishing with a surface grinder presents a problem in selection of an appropriate wheel which will cut the hard materials yet not smear the soft materials, thus loading the wheel. However, there is enough variety in surface grinder wheel construction whereby a suitable wheel can be found by trial and error for any particular combinations of materials. Care must also be taken to insure that the etchants selected for metallographic preparation do not create excessive relief at material interfaces. Again, trial and error can usually provide etchants which adequately display the features to be inspected without adding excessive relief.

### Incremental Grinds

The technique of incrementally grinding, polishing, etching and examining a number of planes of a metallographic sample, in order to get a three-dimensional contour of a feature, is an important tool for evaluation of multi-material products. Sectioning is often done without precise knowledge of the location of the feature of interest; therefore, it is often important to get an estimate of what the feature is like below the surface of the metallographic plan. When data from all the increments are considered together, it is possible to get a three-dimensional picture of the size, shape and orientation of a feature, or the worst-case evaluation of a defect. A representation of an incremental grind is shown in Fig. 1.

### SOME INSPECTABLE ATTRIBUTES

### Internal Dimensions

Multi-material products may often appear externally as homogeneous single-material products. Some materials in the product are often totally encased in other materials to modify the electrochemical (corrosion resistance), mechanical (tensile strength), or nuclear (radiation attenuation) properties of the product. The location, shape, and dimensions of these internal structures often are useful in measuring corrosion resistance (the thickness of cladding over a corrosion-prone material), mechanical properties (fiber length), or radiation attenuation (thickness of attenuating material). Alloys of similar composition must often be distinguished. Figure 2 shows an interface used for dimensional reference where the alloys distinguished are two members of the zircaloy family which only differ by 0.05% Ni. The contrast was obtained using an etchant applied on the polishing wheel and oblique external lighting. Dimensional measurements can be taken most conveniently with a toolmaker's microscope with digital read-out as shown in Fig. 3.

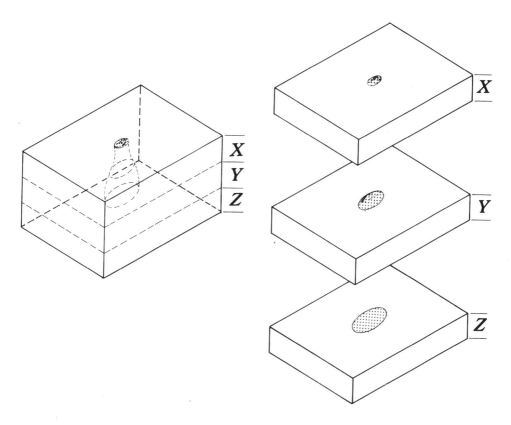

Fig. 1.   Incremental Grinding. Several planes are required to reveal the full indication.

The scope shown provides 4—decimal dimensional accuracy at 35X.   A cross-hair in the eyepiece is lined up with the feature to be measured as shown in Fig. 4.   By zeroing the read-out and traversing to a reference point, internal dimensions can be quickly obtained.

The internal shape of a feature is often important to product quality.   The shape could affect the stress field in the product or the corrosion protection of a plating or cladding in a critical area.   If the shape is one not readily describable by linear dimensional measurement, it is often easier to use an optical comparator with a template to evaluate the shape.   Figure 5 shows a typical comparator with a metal-lographic mount positioned for inspection.   When inspecting features on metallograph-ic surfaces, the light is reflected off the surface of the sample and topographic differ-ences between the two materials cause reflection differences which can be detected on the screen.   In Fig. 6, the curvature of the metal-to-metal interface is important to product quality.   A template with curves which represent the minimum, maximum and nominal curvature at 50X is placed on the viewing screen for comparison.

Thermal History

The thermo-mechanical cycles that a multi-material product experiences are often

Fig. 2.  Interface between Zircalloy 2 and Zircalloy 4.

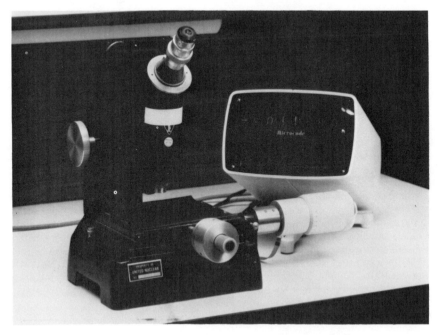

Fig. 3.  Toolmaker microscope with digital readout.

Fig. 4. Toolmaker microscope eyepiece crosshair superimposed on interface.

Fig. 5. Optical comparator.

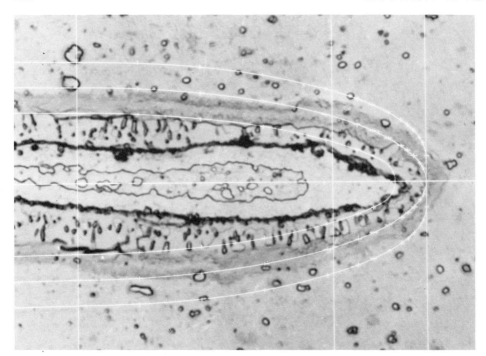

Fig. 6. Template on optical comparator screen superimposes minimum and maximum curvature lines on the feature image to confirm shape acceptability.

important to its quality and performance. Mechanical strength and corrosion resistance are related to the microstructure. Many attributes, such as phase and constituent presence in the microstructure, grain size, and precipitation are typically used to verify thermal and mechanical history in products. In some cases, there are unique sensors within multi-material products which enable more accurate measurements on thermal input. Figure 7 shows a diffusion region where Material A diffuses into Material B at a very consistent known rate as a function of temperature and time-at-temperature. Therefore, the thickness of the diffusion region provides an accurate measurement of annealing time and temperature, and other thermal input. If a multi-material product contains such a diffusion couple, it can be readily used to monitor heat input which might be detrimental to mechanical properties where grain size measurements are too insensitive. If no such couple exists, it may be possible to dope the multi-material product with a diffusable particle of another substance for the sole purpose of thermal input calibration.

Bonding Integrity

Today, various sophisticated techniques are used to produce strong diffusion bonds between dissimilar materials. It is important to inspect these bonds for integrity; foreign inclusions, resulting in bonding defects in the product. Figure 8 shows such a foreign inclusion.

Even contamination so slight as to be invisible metallographically can prevent grain growth across the prior interface. Figure 9 shows another foreign inclusion, but note, that with polarized light, the grain boundaries to the left of the defect are

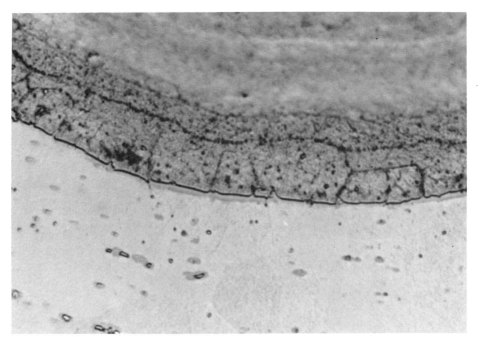

Fig. 7.  Diffusion region between two dissimilar materials.

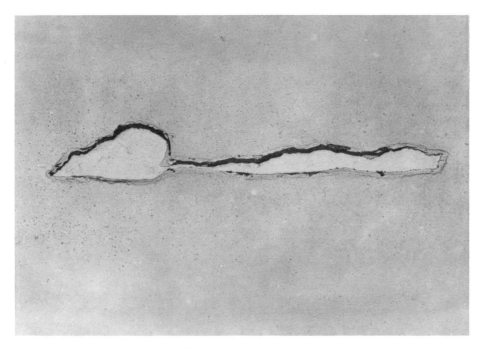

Fig. 8.  Foreign inclusion.

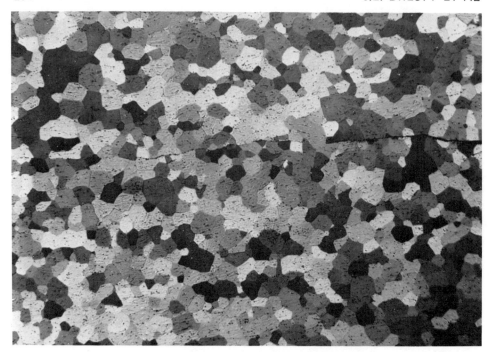

Fig. 9.  Foreign inclusion with coincident grain boundaries.

coincident with the prior interface.  Grain growth has not occurred;  therefore, the
bond is substantially weakened.  The defect in Fig. 9 should be measured to the
end of the coincident grain boundaries for accurate correlation with bond strength.

Powder Product Integrity

Powder metallurgy parts as components in multi-material products depend on
the use of high quality homogeneous powder which is free from contamination.
Metallographic examination of a powder metallurgy part can disclose hard or soft
contamination spots or areas of incomplete densification.  Figure 10 shows such a
hard contamination spot.

Nondestructive Metallographic Testing

Metallography can also be used as a nondestructive testing tool for polishing
and etching multi-material products that will actually be put into service.   The technique has been considered useful in "field metallography", but recent advances in
equipment increase the attractiveness of the procedure as a standard laboratory inspection tool.  A typical device for such inspection (molvipol) is shown in Fig. 11.
By proper choice of electrolyte and voltage, the device can remove plating from the
small nozzle area and expose the substrate metal below.  By further adjustment, the
substrate metal can be electropolished, and still further adjustment achieves electrolytic etching of the surface.  The etched area can then be examined microscopically
either directly or on a replica to determine microstructural features.

Figures 12 and 13 show the results of the technique on Inconel X750 with
AH and BH heat treatment respectively.  Note the fine precipitation of gamma prime

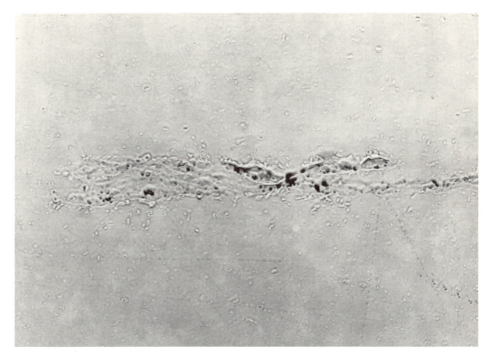

Fig. 10. Hard contamination spot in a powder product.

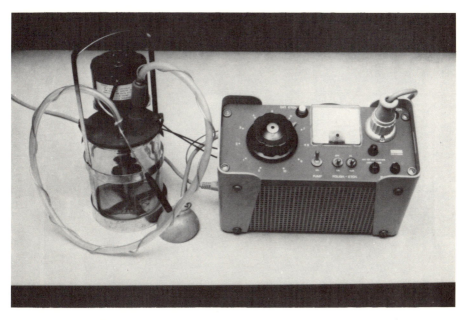

Fig. 11. Equipment used for nondestructive metallographic testing.

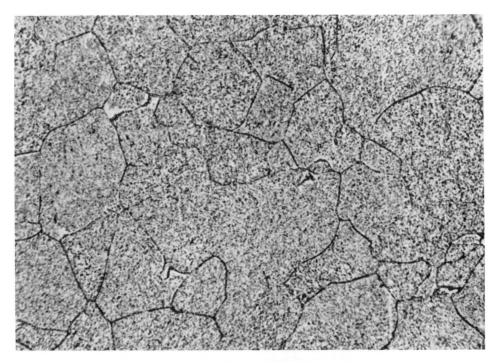

Fig. 12.  Inconel X750 with **AH** heat treatment.

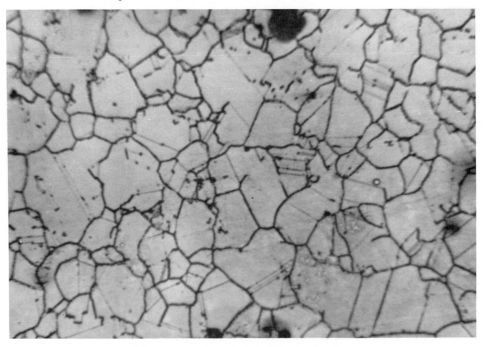

Fig. 13.  Inconel X750 with **BH** heat treatment.

phase found in the AH structure, but absent in the BH. The materials are easily confused due to their similarity in properties and processing, both at the mill and the fabricator's plant. BH has superior properties in some applications therefore, the ability to distinguish the material in finished parts is important.

Unexpected Features

In the metallographic evaluation of multi-material products, metallographers should always be watchful for unexpected features in the microstructure. The potential for occurrence of such unexpected anomalies is greater with multi-material products because of the sensitive manufacturing processes used, and their propensity for contamination. For example, surface oxidation of powder used for a powder metallurgy part may embrittle the material causing microcracks to form during manufacture, as shown in Fig. 14. If such cracks are found, the storage and processing of the powder raw material will be indicated as the potential cause of inferior quality in the finished product.

EVALUATION OF DATA

The data obtained from metallographic destructive evaluations of multi-material products can be evaluated to make a statistical confidence statement concerning the quality of units not destructively evaluated but actually put into service. The procedures involved are standard industrial statistics, however, these techniques are especially relevant to multi-material products where product attributes are related to complex manufacturing process parameters.

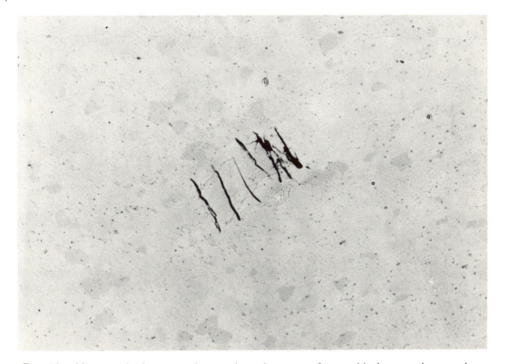

Fig. 14.  Microcracks in a powder product due to surface oxidation on the powder.

Statistical control limits can be established which will assure the quality of production components through destructive evaluation. Though the field of industrial statistics in quality control is a large one, a few salient points can be made.

The essence of any destructive evaluation program is the representativeness of the components sampled. A destructive evaluation selection program must be random. A good method is to group components into batches based on their manufacturing history, then select the particular components to be evaluated using a random number table.

The critical operations which affect each attribute must be determined. Though some critical operations will be obvious, empirical evaluations should be performed to locate all significant sources of component variability. The destructive evaluation should occur soon after the last critical operation to assure timely feedback and minimize the quantity of product in jeopardy. The attributes to be evaluated must be quantifiable and bear a meaningful relationship to the critical operations that control quality and product variability.

Guidelines should be setup on the evaluation of the attribute to minimize subjectivity and eliminate biases between inspectors. A blind crosscheck system with reasonably tight limits of agreement, samples with "assigned" values, detailed qualification requirements, and detailed procedures are established methods for maintaining the accuracy of evaluations over a period of time.

Effective control limits require a detailed knowledge of the particular attributes when the manufacturing process is "in control", i.e., producing acceptable products. To establish this knowledge, a number of components (usually at least 20) are destructively evaluated. The attribute average $(\bar{X})$ and the attribute variability represented by the standard deviation (S) or the average range $(\bar{R})$ are established for the process.

One type of control limits is developed by suppression of the product acceptance limits. The upper (UCL) and lower (LCL) control limits would be:
UCL = Max. − K(S) and LCL = Min. + K(S). K is a statistical confidence factor that can be obtained from standard statistical tables. For most evaluations which require a high degree of assurance, a K−95/99 factor is used. This factor is derived such that when the average value is within these limits at least 95% of the time at least 99% of the components will not exceed the product limits. If the value of an attribute in a sampled component falls outside these limits, it cannot be guaranteed that all components from that group are acceptable.

A preferred method is to establish control limits based on process capabilities. If the value of an attribute in a sampled component falls outside these limits, then this indicates that the process is "out of control", i.e., different from when the original limits were established.

For an attribute which has several readings taken on each component, e.g., plating thickness, both average and range control limits can be used.

If each component (j) is sampled at (i) locations for thickness (T) then: The average thickness for the jth component would be

$$\bar{X}_{T_i} = \frac{1}{n} \sum_{i=1}^{n} T_{ji}$$

The range would be $R_{T_j} = T_j \text{ max.} - T_j \text{ min.}$

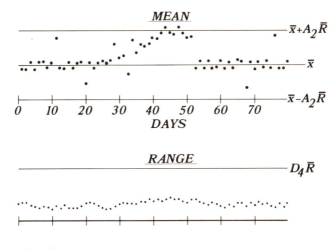

Fig. 15. Control charts for process mean and range.

If many components are evaluated then the process average and range would be:

$$\bar{X}_T = \frac{1}{m} \sum_{j=1}^{m} \bar{X}_{T_j} \quad ; \quad \bar{R}_T = \frac{1}{m} \sum_{j=1}^{m} R_{T_j}$$

Average control charts would monitor each $\bar{X}_{T_j}$ to verify that the process *average*

is in control. Range control charts would monitor each $R_{T_j}$ to verify that the process
cess *variability* is in control. A summary of the control limits is shown below.

The average control limits would be:

$$UCL = \bar{\bar{X}} = A_2 \bar{R}$$
$$LCL = \bar{\bar{X}} = A_2 \bar{R}$$

Where $A_2$ is a confidence factor
from standard statistical tables.

If the value of an attribute ( $\bar{X}_{T_j}$ ) falls within these limits there is 95% confidence that the *average* of the product is not different from the established process
average.

The range control limit would be:

$$UCL = D_4 \bar{R}$$

Where $D_4$ is a confidence factor from
standard statistical tables.

If the range of an attribute ($\bar{R}_{T_j}$) does not exceed this limit, there is 95% confidence that the *variability* within each component is not greater than the established
process variability.

These control limits should be periodically updated based on the most current
knowledge of the process. By establishing control charts, as shown in Fig. 15, with
any of the above control limits, the manufacturing process can be accurately moni-
tored for process shifts and corrective action taken before many components must
be reworked

# HOW METALLOGRAPHY AND SEM ANALYSIS CAN BE USED IN QUALITY CONTROL FOR MICROELECTRONICS

Garry W.E. Johnson*

## INTRODUCTION

Metallography has sometimes been thought of as "Black Magic", but to the innovative engineer the results are actually a combination of both art and science. Success in the use of metallographic techniques in the electronic industry requires several disciplines in mechanical, electrical or chemical engineering to interpret the information that a microsection can provide.

Until recently, metallography has not been routinely used as a technique in the electronic industry as part of a quality control procedure. Because of the continued demand for better reliability and the need to monitor vendor's processes and inspection procedures, new laboratory techniques had to be developed. As these new metallographic techniques were being developed, electronic companies were implementing these ideas in their material evaluation procedures. Therefore, there is an ever growing need for electronic and material engineers to understand each others disciplines and have a closer working relationship. When this happens it will be easier to determine whether failures are vendor related or generated within the company.

Before discussing metallographic and SEM techniques, let's briefly define quality control. Quality control is a systematic process to ensure some level of reliability by testing good parts that might become defective due to mechanical, electrical, or other failures. System failures are shown to occur due to either improper device designs, or inefficient inspections and testing which sacrifice reliability.

In the past, reliability improvements have been investigated only at the final assembly level. Now the main objective is to have a sound quality control program at the beginning, involving the analysis of each part to insure some degree of reliability. The higher reliability desired, the higher the cost. If low cost is the main objective, some degree of reliability will be sacrificed.

On the market place today, quality control test evaluation programs are generally related to either military or commercial products. The military generally demands higher reliability than commercial companies and generally have well planned quality control procedures which can be found in most military specification standards. Commercially, quality control procedures generally vary between companies and follow no set pattern, except that most semiconductor companies follow some form of the military specification standards, but at a low reliability level.

* ISS/SPERRY UNIVAC, Cupertino, California 95014 USA.

Because of the complexity and variations in electrical testing, this subject will only be discussed briefly. There are some very good references given in this paper if more information on this subject is desired.

During incoming inspection of electronic components, quality control procedures follow certain steps to evaluate these components (see Fig. 1). First the components are subjected to some level of electrical testing. This level of testing is directly related to the desired degree of reliability. For example, integrated circuit devices are generally tested by some type of automatic tester. These testers may range from simple to very complicated and can be expensive. Again, the reliability level dictates what testers are to be used.

Mechanical evaluation involves hermiticity, shock test, X-ray analysis, etc. after which most parts can be returned to the main population and used in the system.

In contrast, material evaluation is a destructive physical analysis (DPA); once the parts are tested they cannot be used in the system. Most parts used for this type of analysis are subjected to a complete internal inspection including microsectioning. (The subject of microsectioning will be discussed in detail later).

When the final evaluation has been completed, the data should be documented, then if a problem does occur, this data may be used to help solve the problem.

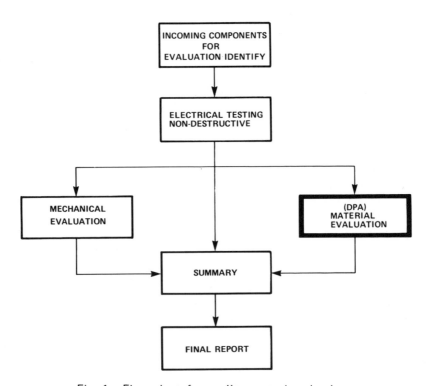

Fig. 1.  Flow chart for quality control evaluation.

## METALLOGRAPHIC ANALYSIS FOR QUALITY CONTROL

Material evaluation procedures as shown in Fig. 2 are a sequence of systematic procedures to determine if a certain electrical component will or has failed because of poor workmanship, latent failures, or other problems related to material defects.

Once the visual inspection has been completed, further electrical evaluation may be required. In some cases, the part may have an unknown contaminant which needs chemical analysis to identify the material. After the above analysis has been completed, a summary should be made to correlate all the information for the next step—microsectioning.

Metallurgical microsectioning is a systematic procedure that must be followed exactly to obtain the necessary information. When implemented correctly, these techniques can be used in a routine manner to examine microelectronic components for quality control.

This discussion will focus on the use of metallurgical microsection techniques for microelectronic components in quality control.

In the past, some microscopists have had difficulties in the electronic industry by trying to microsection a predetermined location in the micron range or less. Because of these difficulties, this part of the material evaluation has sometimes been discontinued. This does not have to happen if a little extra patience and skill are applied to the microsection. Also important is the implementation of a good filing system to document all information found during the evaluation. This information can be very helpful for future reference if a particular problem reoccurs with any one component. The information can then be used by quality control personnel to judge whether the problem is vendor related or in the design of the system.

## PROCEDURE

By reviewing all procedures and techniques prior to polishing as recommended in the references, with some practice and patience, a microscopist should be able to microsection any microelectronic device using the techniques described in this section. Because the parts can be extremely small, pitfalls may be encountered at each step in the procedure. These steps include: taking pictures, encapsulating the devices, coarse and fine grinding, rough and fine polishing, and etching or staining when necessary. But with practice and the use of fast curing epoxies, a device for Quality Control analysis can be microsectioned in several hours.

All techniques require the ultimate in cleanliness and the following of each step carefully. The lack of any scratches is mandatory before one can safely interpret the information from the microsection.

Even though several papers have been written (including reference papers) using diamond abrasives for coarse and fine polishing of an encapsulated electronic component, the author would like to make several comments about why diamond abrasives should not be used when microsectioning electronic components. Diamond abrasives are used in most laboratories and in some cases are a must to achieve the end result. But it has been found that it can harm the semi-conductor material used in electronic components. Silicon for an example, which is the main semiconductor material for an integrated circuit, can be fractured or damaged very easily by diamond abrasive. Especially if the device is etched to delineate the junctions on the polished silicon, the scratches from the diamond particles can propagate in all directions. This is very

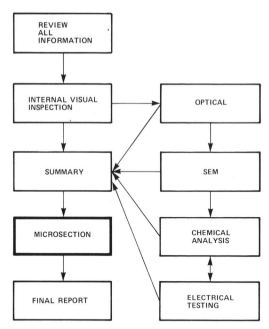

Fig. 2.  Flow chart for material evaluation.

important when trying to microsection an area one micron or less, the etched scratches can cover up the area of interest.  Also if the diamond abrasives are not properly used and there are poor cleanliness habits in the laboratory, the polishing wheels can get cross-contaminated causing serious problems during the polishing procedures.

The following metallographic procedures, if followed correctly, can be used for microsectioning most types of electronic components such as diodes, integrated circuits, etc.  To best illustrate these procedures, an integrated circuit will be used as an example (see Figs. 3–8) for microsectioning.  Then after describing the procedure, typical illustrations will be given on how to apply these techniques to quality control.

Step 1 —
After the device has been evaluated and has been properly opened (if necessary), pictures must be taken of the area of interest so the microscopist will be familiar with the area to be microsectioned (see Fig. 3).  Once all information given is understood, the process to microsection a device can begin.
Step 2 —
The device to be microsectioned has to be encapsulated in some type of medium because most electronic components are generally too small and fragile to handle independently.  The most important fact is that because of their fragility, most components cannot take any pressure during the mounting procedure.  That is why the encapsulating medium should be selected on the following bases.
1. The medium should be transparent for observation of the device during the microsectioning procedure.
2. Hard material is essential to prevent any rounding at the edge of the device.
3. No shrinkage can be allowed to occur at the edge of the device.  This is to prevent voiding.  The encapsulating material has to come in complete contact with the device.

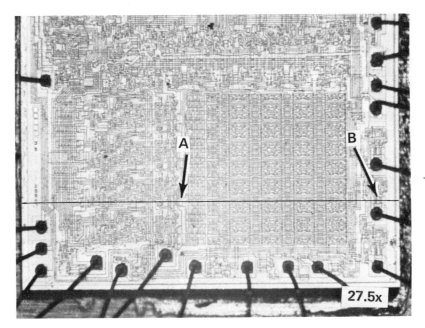

Fig. 3. The surface topology of the integrated circuit to be microsectioned. (Arrows point to the area of interest).

4. The medium has to be low enough in viscosity to allow trapped air bubbles to escape during the curing time.
5. In order to prevent 'pseudo' damage to the device, curing temperature must be kept to a minimum. Most epoxies generate their own heat (of about $80^\circ C$ which is considered alright) during the curing process.

The two epoxies the author has found having the best qualities mentioned above are:

Epoquick — Epoxy manufactured by Buehler Inc.
Shell Epon 815 and Catalyst "A"

Epoquick epoxy is becoming more popular all the time because of the fast curing time (about one hour) at room temperature ($25^\circ C$). Some bubbles do get trapped in the media and are difficult to evacuate by vacuum. If air bubbles are a concern, the Shell Epon is a good epoxy, but curing time takes at least eight hours at $70^\circ C$. The air bubbles can be eliminated very easily by a vacuum system.

Encapsulating Technique

Several major concerns must be considered before encapsulating process begins. First, be sure the part is cleaned of any foreign particles that might have been introduced during the handling. Normal laboratory techniques for cleaning semiconductor devices are sufficient. Secondly, if the part is packaged in a ceramic material, this material will have to be removed before the part can be microsectioned. This is because the ceramic material is so hard it will tear the polishing cloth. There are several techniques that can be used to remove the ceramic. If it is an IC, the device can be placed on a hot plate and heated to melt the solder material holding the IC in the package, this technique can sometimes do harm to the device if not used properly.

The other technique is to encapsulate the whole package, grind the ceramic off with silicon carbide papers, and remount the part in the epoxy. Care has to be taken as not to damage the device with the papers. The last resort would be to purchase a machine to remove the die from the package with little or no damage to the device. There are several of these machines on the market but they can be expensive.

Some type of specimen holder has to be used to hold the media and device while the epoxy is curing. There are several companies that sell these. Or you can make your own out of some type of plastic material. (Be sure to use a mold release or the epoxy may stick to the specimen holder).

Once the above problems are worked out, the part can be encapsulated (see Fig. 4). Mixing and curing the epoxy should be done by following the recommendation of the manufacturer; if any serious problems do occur, contact your local representative who carries metallurgical supplies.

Once the encapsulating material has properly cured, the mount can be removed from the holder and properly identified. Now the next step may begin.

Step 3 — Viewing Window

The author has found that one of the most important aspects of preparing the mount for microsectioning is placing a polished flat surface on one side of the mount (see Fig. 5). This polished flat surface should be about one-quarter of an inch from the device. When micro-sectioning an area of several microns or less, this technique will enable the microscopist to view the mount either with a low power or high power microscope. A good procedure for doing this is to grind the epoxy with 80 or 120 grit silicon carbide paper, next use 400 grit to remove the previous damage and final polish using Linde "A"* (0.3) microns) on a silk cloth.

After the viewing window has been placed in the right orientation, the mount should be ready for rough and fine grinding.

Step 4 —

The rough and fine grinding operation generally takes several steps with different grit sizes of silicon carbide papers. A cardinal rule is never touch any part of a semiconductor material with silicon carbide paper larger than 400 grit as this will permanently damage the material. Experience indicates the most efficient way of grinding and polishing is holding the mount with cut view window towards the operator. It is then placed firmly on the silicon carbide papers with water flowing to prevent the sample from over heating. Never rotate the sample around the periphery of the wheel (this is contrary to normal polishing techniques) (see Fig. 6). It has been the experience of the author that the elements that make up the device can be damaged very easily by rotating the sample on the wheel. The best way is to move the sample firmly with medium pressure back and forth from the center of the wheel to the outer edge, occasionally viewing the mount through the viewing window with a high power magnification. Also since a sharp edge is required at all times, it is a good idea to change the papers occasionally. Leaving the papers on the wheel too long is a common mistake made by some operators, which can only do more harm than good.

Following the above procedure, the grinding and polishing operation can best be described below (see Fig. 7).

---

*AL$_2$O$_3$ polishing abrasive purchased from Buehler Corp., Evanston, Illinois USA.

Fig. 4. The technique for encapsulating a part in epoxy.

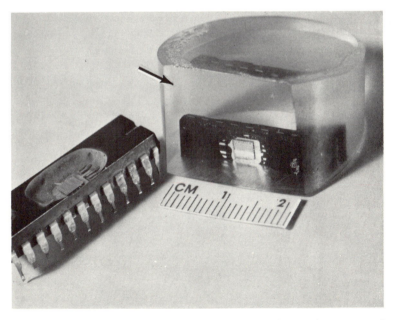

Fig. 5. A device that was encapsulated in epoxy with a flat surface, cut and polished on one side of the mount. This enables the microscopist to view the device when microsectioning. Device on left is same type before being mounted.

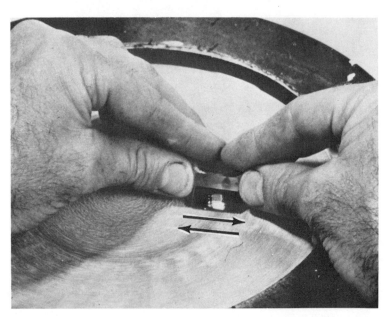

Fig. 6.  The technique used to hold the mount when microsectioning (see arrows).

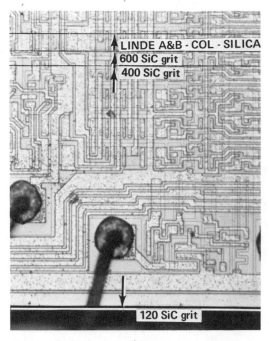

Fig. 7.  The different steps and where to stop at certain points when grinding and polishing a device.

- Rough Grinding — This step is for removing the encapsulate material up to the device using 80 or 120 grit silicon carbide paper. Make sure never to touch the device as this will shatter the material.
- Fine Grinding — This is generally a two step operation first using 400 and 600 grit paper. It would be safe to grind to within .002 or .003 of an inch to the area of interest. Be very careful and view sample occasionally as one mistake can ruin the sample forever. Finally it can be taken to within .0005 of an inch of the target area using 600 grit paper. Once this has been successfully done, wash the sample with soap and water before going to the next step, which is the rough and fine polish procedure.

Step 5 —

This step will conclude the final process for microsectioning an electronic device. If all previous steps were followed correctly, the final polishing should be completed without much difficulty. Here again, the technique for holding the sample is the same as the previous operation (see Fig. 6). Remember that cleanliness is very important.

- Rough Polishing — This is the most critical step of the entire grinding and polishing procedure. Not only do all the previous abrasive scratches have to be removed, the polishing has to stop right on the area of interest. One small miscalculation can ruin the sample. The polishing abrasive is Linde "A" suspended in DI water. The best container is a plastic squeeze bottle with two or three spoonfuls of abrasive, filled to the top with DI water (they can also be purchased from Buehler already mixed). The author has found the silk cloth on a Buehler brass wheel (ring type) cuts better and faster than any other cloth on the market. The polishing wheel should be about medium speed and during the polishing operation occasional squirts of abrasive are necessary. Also a lot of pressure is sometimes needed to cut into the final area of interest. This depends mainly on how large the sample is being microsectioned. Care has to be taken to be sure not to polish past the area of interest. Cutting can happen quickly if not watched carefully.
- Fine Polish — This step is used to remove the Linde "A" scratches (not much cutting action happens in this step). The cloth used is a medium nap Microcloth*. It is ring mounted on an 8 inch brass wheel using Linde "B" (.05 micron) as the abrasive. Again cleanliness of the sample and polishing cloth is critical. The polishing wheel should be low to medium speed and with medium hand pressure by the operator. Several squirts of abrasive should charge the cloth enough to polish one mount. This step should last not more than sixty seconds, otherwise severe rounding of the sample will occur. After the Linde "B" polish, there still can be polishing scratches left on the surface that might interfere with the interpretation of the microsection. So the author has introduced an additional step called the final polishing. It leaves a scratch-free surface which is very important when microsectioning in the submicron range.
- Final Polish — Cleanliness is essential in this step. The polishing cloth used is Pix Cloth** using colloidal silica† as the abrasive. Colloidal silica has to be kept moist at all times and completely cleaned off the polishing cloth after use. If it dries on the cloth it will become hard and will scratch the sample easily. Use the same polishing technique as the previous steps, and

---

*Buehler Ltd., Evanston, Illinois USA.
**Manufactured by Geoscience, Mount Vernon, N.Y. USA.
†A product of PQ Corporation, Lafayette Hill, PA USA.

pour from the bottle just enough, as needed. Rotate the sample back and forth for 60 seconds or less. When finished, wash off immediately. Let the water run on the cloth, distilled water if possible. If all these steps are followed properly, the results can be very rewarding. The author has implemented this step in the normal polishing procedures in the metallography laboratory.

Step 6 —

Most semiconductor devices need to be etched or stained in order to delineate the junctions after a component has been microsectioned (see Fig. 8 A,B). Because there are many techniques and etches that can be used, this paper will not cover this area.

Step 7 —

After the above procedures have been finished, pictures are now taken of the area of interest (see Fig. 8 A,B). This step generally calls for using either a horizontal or vertical metallograph and to be able to take pictures with magnification up to 3000X. But with areas of interest in the submicron range, the SEM can be used. The epoxy mount is a non-conducting material so a thin layer of gold or carbon has to be sputtered on the surface to make the surface conductive before the mount can be used in the SEM.

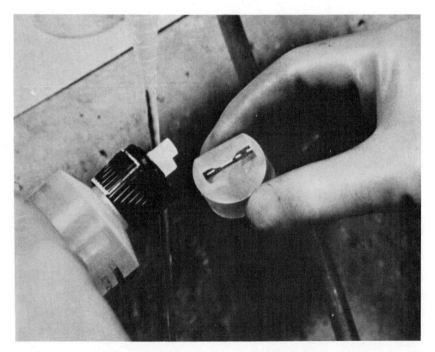

Fig. 8. The technique used for etching a microelectronic device. This process has to be done in an exhaust fume hood and with the right type of laboratory clothing worn.

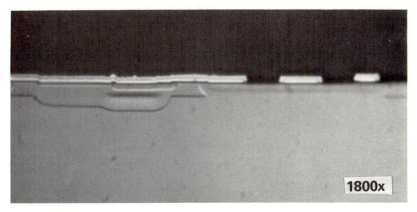

Fig. 8A. An area microsectioned in conjunction to surface topology photograph shown in Fig. 3.

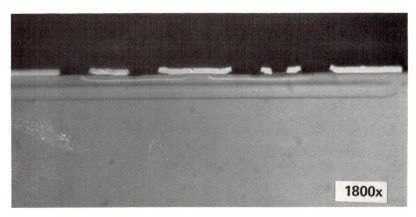

Fig. 8B. A MOS transistor microsectioned in conjunction with surface topology photograph in Fig. 3.

Illustrated Examples

To illustrate the techniques just discussed, the following examples demonstrate how microelectronic components can be observed internally for quality control. Included are examples of scanning electron microscopic (SEM) and metallographic techniques used to help evaluate an integrated circuit that was found to be poorly fabricated.

1.  Examples are shown in Fig. 9 of two solid tantalum electrolytic capacitors that were microsectioned to determine whether or not the internal construction meets the specifications. Both parts passed the electrical tests, but one of the parts (B) failed because of internal material defects.

    In Figs. 10 and 11, the microsections show the internal construction to pass material specifications. (Note in Fig. 11 how the final polish can illustrate the different layers without etching the sample).

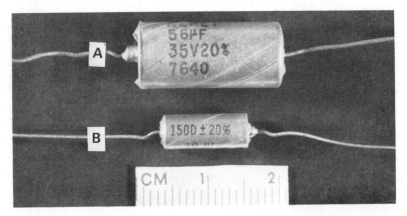

Fig. 9. Two capacitors that were microsectioned to illustrate how metallographic techniques can identify defective parts for quality control. Figures 10 through 13 show the same type of parts microsectioned.

Fig. 10. A metallurgical microsectioned solid tantalum electrolytic capacitor which shows the internal construction.

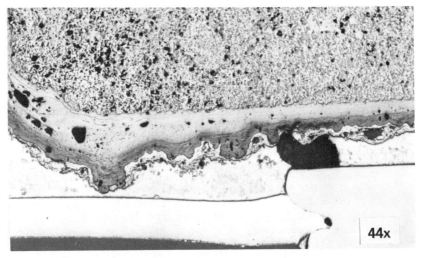

Fig. 11. A portion of the capacitor shown in Fig. 10 at a higher magnification.

Even though the part shown in Fig. 9 had passed electrically, the micro-sections in Figs. 12 — 13 show the internal construction to be of poor quality and does not meet the material specifications. Because of this poor construction the part could fail electrically at any given time. It appears that movement inside the device had occurred when fabricated. Two other capacitors of the same type were microsectioned and showed similar results. These results were shown to the vendor and corrective action was taken.

2.   As the microelectronic industry is rapidly expanding into the submicron range, the need for instrumentation to produce better resolution at higher magnifications is a must. The following example is a case history that required this higher resolution instrumentation to solve a problem that occurred in an integrated circuit (IC) that is used on an arm assembly for the disk drives (see Fig. 14). These disk drives are part of the mem-ory units for the computer systems. This particular IC was failing at both the incoming inspection level (QC) and out in the field. This example will show how the scanning electron microscope (SEM) and metallographic techniques were both used to help solve the problem.

The parts that failed in the field and at the incoming inspection level (QC) were first delidded and examined with the optical microscope. It was found that the devices could not be examined optically at high mag-nification because they were too small and the IC package would inter-fere with the objectives (See Figs. 15 and 16). Therefore, the SEM had to be used for most of the investigation. As can be seen in Figs. 17 and 18, the device appeared to have failed where the double metallization traces overlapped each other. (This is where two metallization (aluminum) traces are on top of each other with a dielectric material between the two). After further investigation of the same devices with the SEM, it appeared that the top layer of metallization had not been properly removed when fabricated (see Fig. 19). To better see this, the glassivation layer (top layer $SiO_2$) was etched and removed with hydrofluoric acid. This process exposed finger-like metallization protruding away from the main trace down to the bottom layer of metallization (see Fig. 20).

Further investigation into other devices revealed the same characteristics. Some of these failed devices were mounted in epoxy and microsectioned in the areas of concern. As can be seen in Fig. 21, the failure had oc-curred at the metallization traces. Other failed devices were also micro-sectioned, revealing the same characteristics. Also, areas that showed these finger-like metallization traces were microsectioned. These can be seen in Figs. 22, 23, and 24 which show the top metal traces coming very close to the bottom traces. Several things can happen when this oc-curs. If there are pin holes in the dielectric (protective layer to keep the two traces from shorting out) metallization can migrate through these holes and cause the failure. Secondly, if the metal traces are too close to one another, they can fail electrically. Note Fig. 23 which shows metal in the process of migrating through a pin hole. As can be seen in Fig. 24, the glassivation layer can be seen very well in the microsection. It also reveals a good coverage of $SiO_2$ on top of the metallization but not at the edges. This can be important if contamination is on the sur-face. Many other devices were examined with this same process, the re-sults were all similar. This information was presented to the vendor who improved the yield and tightened controls at the quality control level. This included better burn-in testing and improved visual inspection.

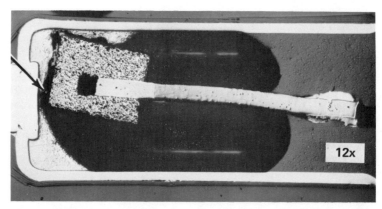

Fig. 12.  A metallurgical microsectioned solid tantalum electrolytic capacitor which failed the specification.

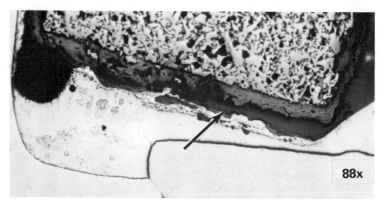

Fig. 13.  A portion of the capacitor shown in Fig. 12 at a higher magnificatio Note the tantalum slug which pulled away from the inner wall.

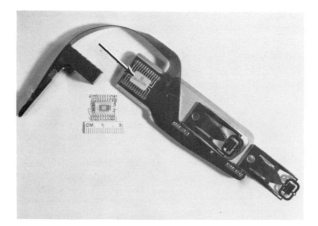

Fig. 14.  An arm assembly with the integrated circuit soldered to it (see arrow). The IC shown with scale is the same type mounted on assembly arm.  The top lid has been removed.

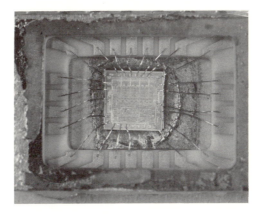

Fig. 15. A higher magnification of the part shown in Fig. 14 which illustrates how the silicon chip is mounted in package.

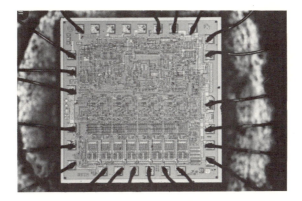

Fig. 16. The device shown in Fig. 15 but at a higher magnification (optical photograph).

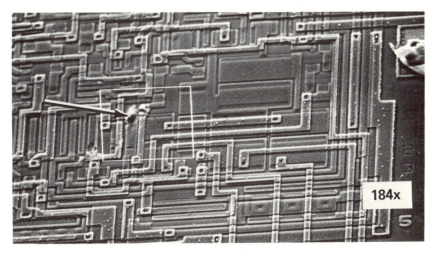

Fig. 17. Same type of IC that failed in the field (see arrow).

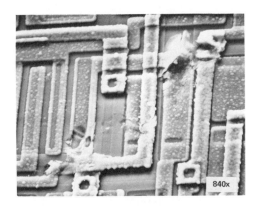

Fig. 18. Same failure as in Fig. 17 but at a higher magnification.
(Reduced 45% for reproduction.)

Fig. 19. Metallization that wasn't properly removed (see arrows).
(Reduced 45% for reproduction.)

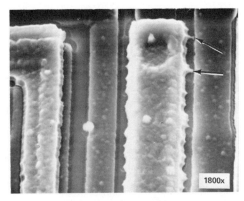

Fig. 20. The same IC as in Fig. 19 but the glassivation has been removed. Note finger-like metallization (see arrows). (Reduced 45% for reproduction.)

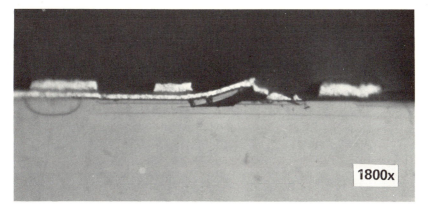

Fig. 21.  A microsection of the failed IC.

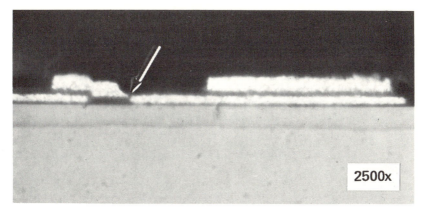

Fig. 22.  A microsection of the finger-like metallization protruding from the top layer to the bottom.

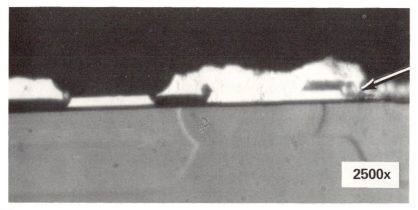

Fig. 23.  A microsection of the finger-like metallization and metal migrating through a pin hole.  (See arrow).

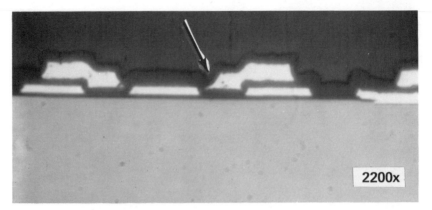

Fig. 24. Finger-like metallization and shows the top protective layer of SiO$_2$ (glass-ivation layer). Note how thin the SiO$_2$ is on the walls of the metallization (see arrow).

## CONCLUSION

The most important aspect of quality control is trying to determine if a part will or will not fail in a system due to poor quality. When a failure does occur, is the poor workmanship vendor related or generated within the company? This discussion has hopefully answered some of these questions by introducing the concept of using metallographic microsectioning techniques to determine the quality of a given part.

If the procedures as shown in this paper are followed correctly, the microscopist should be able to microsection any type of microelectronic component in a short period of time. But most importantly the Material and Electronic Engineer must work together in solving these problems or the effort to bring better quality to a product will be fruitless.

## ACKNOWLEDGEMENTS

The author thanks Peter Mee for his assistance in reviewing this paper for publication. Many thanks to D. Tonick and B. Ream of the ISS/Sperry Univac Art Department for their assistance in the art work and photographs. Also thanks to L. Olmstead for her patience in typing this paper.

## REFERENCES

1. John R. Devaney, "Precision Metallography of Microelectronic Devices", Hi Rel Laboratories Inc., (1978).
2. R.F. Cunningham and D.A. Clifford, "Sample Preparation Techniques for Optical and Electron Beam Analysis", Rockwell Int. Corp., (1977).
3. Robert Anstead, "Techniques of Final Preseal Visual Inspection", Goddard Space Flight Center, (1975).
4. Steven S. Baird, "Failure Analysis", Texas Instruments Inc., (1969).
5. William C. Coons, "The Role of Metallography in the Analysis of Failures of Electronic Components," Lockheed Missiles & Space Co., (1965).

# EFFECT OF REVERTED AUSTENITE ON THE MECHANICAL PROPERTIES AND TOUGHNESS OF A HIGH STRENGTH MARAGING STAINLESS STEEL CUSTOM 450®

J. Edwin Bridge, Jr. and Gunvant N. Maniar*

## INTRODUCTION

One of the problems confronted by a metallurgist, whether in Quality Assurance or Research, is that of the control of austenite in ferritic or martensitic steels. The martensitic transformation in many alloys does not reach 100% completion with the structure containing some amount of "retained Austenite". Similarly during aging of the maraging stainless steel alloys, "reverted austenite" is formed concurrent with the precipitation of an age-hardening phase. The amount of austenite, retained or reverted, has a marked effect on the properties of these alloys. It should be also mentioned, that in austenitic steels, the formation of delta ferrite during high temperature anneals and the pseudo-martensite formed during cold working, has a marked effect on properties. Therefore, it is desirable to have a means to rapidly and accurately measure the quantity of austenite for quality control and for research problems.

This paper describes the inter-relationship of mechanical properties, and microstructure of a maraging stainless steel, Custom 450, as affected by austenite reversion. It is shown that this alloy can be aged using a single step treatment, to obtain a unique combination of properties; such as high strength similar to Type 410 stainless steel and corrosion resistance equivalent to Type 304 stainless steel.

The major alloying elements, chromium, nickel, copper, molybdenum and columbium are balanced to produce a martensitic alloy for an optimum compromise between strength, toughness, fracture toughness and corrosion resistance. On cooling to ambient temperature from the annealing treatment, the alloy completely transforms to martensite. The maraging, age hardening reactions occur in a temperature range of 800 to 1150°F with the maximum hardness attainable in the range of 900°F. Aging at 900°F results in a yield strength of 180 ksi. Overaging results in the alloy being most ductile.

The overaging reaction in this and similar alloys, such as PH 13-8Mo, 17-4PH, etc. is accompanied by austenite reversion. The effect of reverted austenite on the mechanical properties of high strength maraging steels is somewhat contradictory. This study shows Custom 450 is strengthened by the precipitation of Laves phase, whereas the mechanical properties and corrosion resistance are dependent on the volume fraction of reverted austenite.

® Registered Trademark of Carpenter Technology Corporation.
* Research and Development Center, Carpenter Technology Corporation, Reading, Pennsylvania 19603 USA.

The alloy is very stable as the age hardening Laves phase resists coarsening and growth, and the reverted austenite is unaffected by exposures at cryogenic temperatures. However, the reverted austenite can be transformed by cold working with a significant increase in hardness.

## EXPERIMENTAL PROCEDURE

The material for this study was obtained from a production arc heat of Custom 450 as a 3-inch round bar. The bar was in the annealed and centerless ground condition. Chemical analysis is given in Table I.

### TABLE I
### Chemical analysis of Custom 450R

| C | Mn | Si | P | S | Cr | Ni | Mo | Cu | Co | Cb |
|-------|------|------|-------|-------|-------|------|------|------|------|------|
| 0.032 | 0.34 | 0.23 | 0.029 | 0.008 | 14.96 | 6.52 | 0.77 | 1.51 | 0.16 | 0.68 |

Longitudinal specimens for tensile and Charpy V-notch (CVN) testing were annealed for ½ hour at 1900°F followed by water quenching. The test specimens were machined from these annealed blanks and aged at 900°F, 1000°F and 1050°F for times of 4 and 8 hours, air cooled. The mechanical property data are given in Table II. Transverse data were also obtained and found to exhibit trends similar to the ones for longitudinal data and therefore are not included here.

### TABLE II
### Tensile properties of Custom 450R

| Age Hardening Treatment* °F | | 0.2% Y.S. ksi | Longitudinal UTS | El Percent | RA Percent |
|---|---|---|---|---|---|
| 900°-4 | hours | 182 | 190 | 15.9 | 59.7 |
| " | " | 184 | 192 | 15.4 | 59.7 |
| 900°-8 | " | 185 | 193 | 15.3 | 59.7 |
| " | " | 184 | 191 | 13.9 | 59.7 |
| 1000°-4 | " | 173 | 175 | 14.4 | 63.3 |
| " | " | 172 | 175 | 15.9 | 64.6 |
| 1000°-8 | " | 167 | 167 | 17.3 | 65.5 |
| " | " | 167 | 167 | 17.8 | 65.6 |
| 1050°-4 | " | 154 | 158 | 18.4 | 67.1 |
| " | " | 154 | 158 | 18.5 | 67.1 |
| 1050°-8 | " | 144 | 151 | 20.0 | 67.1 |
| " | " | 145 | 153 | 18.3 | 67.3 |

* All specimens annealed at 1900°F/1/2 h/WQ.

For fracture toughness testing, $K_{Ic}$, compact tension specimens, 1T, and 1″ Bend specimens were rough machined from the bar, annealed at 1900°F/½hour/WQ, finish machined, and aged for four hours at 900°F, 950°F, 1000°F, 1050°F, 1100°F, and 1150°F. The specimens were precracked and fractured in accordance with ASTM E−399. The $K_{Ic}$ values are given in Table III. (Note: In this paper, the term "fracture energy" is used to describe the Charpy V-notch impact test results, rather than the term "toughness" to differentiate from the plain strain fracture toughness ($K_{Ic}$) test results.)

## TABLE III

Charpy V-notch impact and plain strain fracture toughness tests — Custom 450

| Age Hardening Treatment* | Fracture Energy ft-lbs | $K_{Ic}$ [1] Ksi $\sqrt{\text{in.}}$ |
|---|---|---|
| 900°F - 4 hours | 26 | 62 |
| " | 32 | 63 |
| " | 32 | 65 |
| 900°F - 8 hours | 29 | |
| " | 29 | |
| " | 29 | |
| 1000°F - 4 hr | 60 | 74 |
| " | 64 | 76 |
| " | 64 | 78 |
| 1000°F 8 hr | 74 | |
| " | 73 | |
| " | 80 | |
| 1050°F - 4 hr | 88 | 99 |
| " | 89 | 100 |
| " | 93 | 103 |
| 1050°F - 8 hr | 98 | |
| " | 96 | |
| " | 102 | |

\* All specimens annealed at 1900°F/1/2 h/WQ.

(1)  Compact Tension Specimens;  All values
     meet validity requirements of ASTM E-399

Specimens for age-hardening response and reverted austenite studies were aged from 900°F to 1150°F for times varying from 4 to 64 hours, and one for 200 hours at 1000°F. Room temperature hardness measurements were made on all samples to characterize the aging response.

The reverted austenite measurements were made using an X-ray diffraction integrated intensity technique employing a semi-automated diffractometer. The procedure is fully described in Appendix I.

The accuracy of this method is dependent on the amount of preferred orientation exhibited by the austenite and martensite. If the phases are nearly randomly oriented the accuracy can be ± 3% of the amount present. The accuracy was determined using a NBS standard SRM485, see Appendix II. The effect of preferred orientation can effect the accuracy considerably. However, the procedure described in Appendix I corrects the integrated intensity values by an averaging method. An example of this correction for a sample exhibiting a high degree of preferred orientation is given in Appendix II.

The accuracy of the measurement when the phases are oriented was determined to be in the range of ± 30%, which means for a sample containing 5% austenite, the actual value could range from 3.5 to 6.5%. The precision of the method is ± 3% of the amount present.

The precision and accuracy figures must be used with caution, especially whenever preferred orientation is present. Only experience can dictate whether or not the austenite value is highly reliable, but in most cases no other method can be used.

A selected few samples were examined by light and electron metallography to characterize the structure as a function of heat treatment. For light microscopy, specimens were etched in a mixture of 400 ml $H_2O$, 400 ml ethyl alcohol, 400 ml methanol, 400 ml HCl, 20 ml $HNO_3$, 28 gm $FeCl_3$ and 8 gm $CuCl_2$. For electron metallography, the specimens were etched in glyceregia and examined for structure as well as electron diffraction analysis of extracted particles[4].

The transmission electron microscopy (TEM) thin foils were prepared by cutting 20 mil wafers and grinding to 4 mils. Necessary precautions were taken to avoid deformation and heating of foils. The foils were electrolytically thinned in chromic: acetic acid mixture using a modified window technique[4].

## RESULTS

The effect of reverted austenite on the mechanical properties are shown in Fig. 1. Figure 1a shows the age-hardening response of the alloy, hardness and reverted austenite formation vs. aging temperature for 4 and 8 hours. The maximum hardness occurs after aging at $900°F$ and the hardness decreases with higher aging temperatures. The loss in hardness is accompanied by an increase in reverted austenite.

The yield strength and fracture energy (CVN) behavior as a function of aging temperature (4 and 8 hour) are shown in Fig. 1b. The yield strength decreases whereas the fracture energy increases as the aging temperature increases. The difference in aging times, vs. 8 hours, has only a slight effect on yield and ultimate tensile strength, and fracture energy. It can be observed that the differences between the 4 and 8 hour age increases with temperature, but sufficient data are not available to define a conclusive trend. As will be shown later, the differences in mechanical properties, 4 vs. 8 hour age, parallels an equivalent change in the amount of reverted austenite.

Figure 2a shows the mechanical property data as a function of reverted austenite formation. It is evident that the yield strength decreases, the fracture energy increases, the fracture toughness increases with increasing amounts of reverted austenite. The fracture toughness data were obtained only in specimens aged for four hours. These plots give evidence that the differences in yield strengths and fracture energies between the 4 and 8 hour aging times are directly related to the amount of reverted austenite. The inter-dependency of these properties can be clearly seen in Fig. 2b.

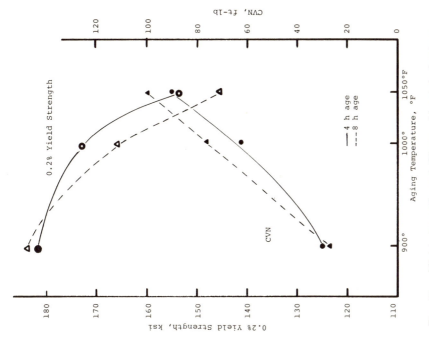

Fig. 1b.  Yield strength and Charpy V-notch fracture energy as a function of aging temperature.

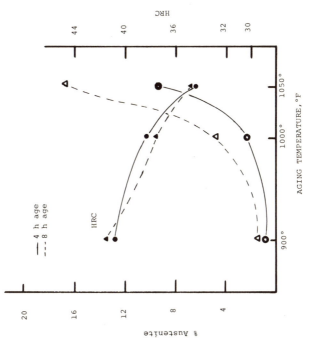

Fig. 1a.  Reverted austenite and hardness as a function of aging temperature.

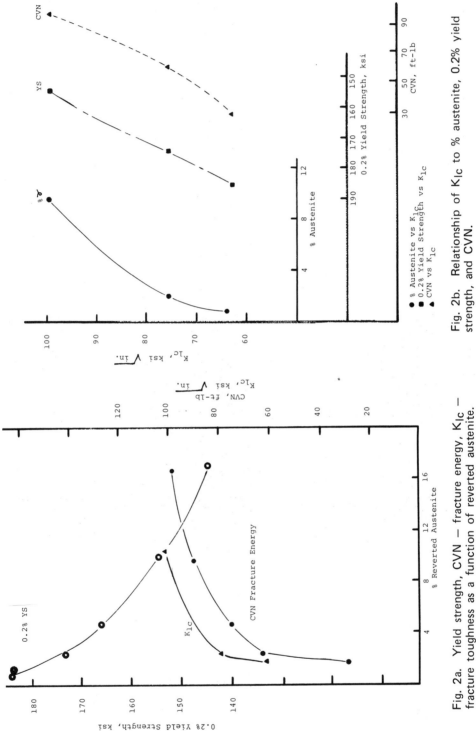

Fig. 2b.  Relationship of $K_{Ic}$ to % austenite, 0.2% yield strength, and CVN.

Fig. 2a.  Yield strength, CVN — fracture energy, $K_{Ic}$ — fracture toughness as a function of reverted austenite.

This is further supported by the fact that the hardness differences between the 4 and 8 hour aging times are not significant, viz., HRC 41/42 at 900°F decreasing to HRC 34/35 at 1050°F.

## MICROSTRUCTURE

The light microstructure of annealed, aged and overaged specimens of Custom 450 were examined. The grain size of the annealed specimen is ASTM 7/8. The structure is typical of low carbon maraging steel.

The age-hardening phase and structures were characterized via electron microscopy. The age-hardening precipitate could not be delineated in structure replicas from the 900°F/4 hour aged sample. The electron diffraction patterns from the extracted precipitates from the 900°F/4 hour specimen showed a diffused halo. An extraction replica was made on the 1050°F/8 hour aged sample with similar results. This indicates that the precipitate particle size is extremely fine, and does not coarsen appreciably even at the highest aging temperature, 1050°F, evaluated for mechanical properties.

### Further Aging Studies

The observations made earlier indicate that the precipitate is ultra fine and highly stable. Therefore, a few specimens were overaged in the range of 900°F to 1150°F for times up to 64 hours to identify the precipitate and study its coarsening rate. This procedure of identifying the precipitate in specimens aged for an extended time is acceptable as long as it is ascertained that a phase change does not occur. This was accomplished by relating the diffraction pattern back to the diffused pattern from short aging time specimens. The effects of these overaging treatments on reverted austenite and hardness were also measured.

The reverted austenite formed in these long aging time temperature specimens and the short aging time specimens was used to construct a temperature-time diagram for reverted austenite formation, Fig. 3, with contour lines constructed for austenite levels. This diagram can be used to determine the rate of austenite formation and for predicting or adjusting the mechanical properties of this alloy for a particular application. Also, the stability of the reverted austenite would have a particular influence on the mechanical properties. Specimens containing 10% and 40% austenite were cold treated in liquid nitrogen for 2 and 64 hours. The liquid nitrogen treatment had no effect on the reverted austenite content, indicating the phase stabilizes rapidly at room temperature.

The behavior of the reverted austenite was further studied by cold rolling material that had been aged for 4 hours from 900°F to 1150°F. Reduction of 10%, 50% and 90% were made and the reverted austenite and the hardness were measured. Figure 4 shows that cold working will transform the reverted austenite with a significant increase in hardness, a behavior similar to that observed in austenitic stainless steels. It is interesting to note that regardless of aging temperature and % reverted austenite formed on aging, the hardness after 90% cold working is ∼ 47 HRC.

The examination of overaged specimens was designed to study the precipitate size, morphology, coarsening rate and to identify the same. From the previous information, the 64 hour aged samples were selected for extensive studies via electron microscopy. The electron micrographs from various overaged specimens are shown in Fig. 5. The precipitate in the 900°F/64 hour age, Fig. 5, is extremely fine. The precipitate coarsened slightly as the temperature increased from 900°F to 1050°F,

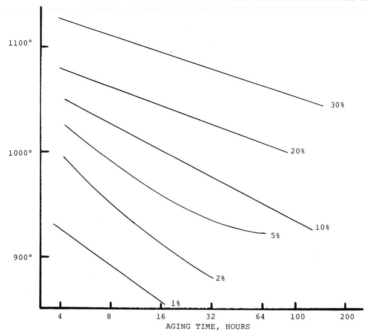

Fig. 3.  Temperature-time diagram for formation of reverted austenite in Custom 450.

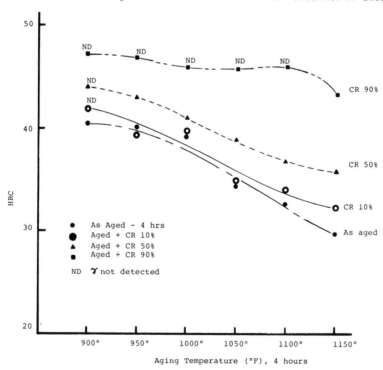

Fig. 4.  Effect of cold working on reverted austenite.

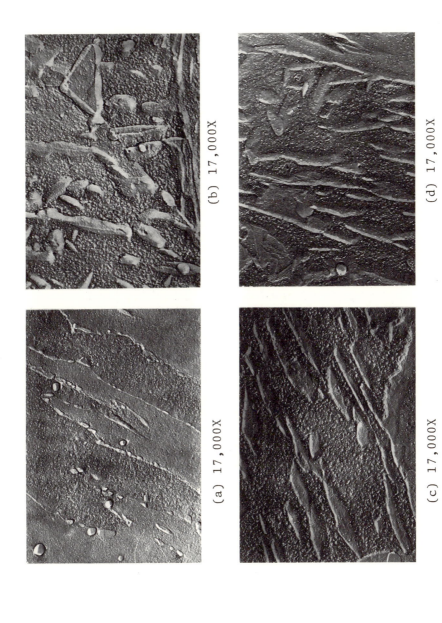

(a) 17,000X

(b) 17,000X

(c) 17,000X

(d) 17,000X

Fig. 5. Electron micrographs of Custom 450 showing precipitate size and reverted austenite. (a) 900°F — 64 hours (2% austenite) (b) 1000°F - 64 hours (19% austenite) (c) 1050° — 64 hours (28% austenite) (d) 1000°F — 200 hours (25% austenite). All solution treated at 1900°F/½ hr./WQ. Chromium-shadowed Parlodion replica. (Reduced 20% for reproduction.)

compare Figs. 5a, 5b and 5c. The $1000°F/200$ hour specimen shows a precipitate size similar to $1050°F/64$ hour specimen, Fig. 5d. Also seen in these micrographs are the ribbon-like reverted austenite formed on the martensite lath boundaries. The increased amount of the austenite for the first three treatments in Fig. 5, from 2 to 28%, is self-evident. The specimen in Fig. 5d had a similar amount of reverted austenite as the specimen in Fig. 5c.

Extraction replica electron micrographs from the same series of samples are shown in Fig. 6. The electron diffraction patterns made from the extracted particles in these replicas, Fig. 7, became sharper with the increasing precipitate size. The precipitate was identified as Laves phase, $A_2B$ type, where the B atom has a larger radius than that of the A atom. In the case of Custom 450, the B is niobium with some molybdenum and the A can be either iron or chromium.

The morphology and actual size of the precipitate and its relationship to the matrix is of particular importance. Thin foils from $900°F$, $1000°F$ and $1050°F$ treatments were prepared for transmission electron microscopy. The electron micrographs in Fig. 8 show the fine dispersion of the Laves phase. The size change from 4 to 64 hours at $900°F$ is very slight and electron diffraction patterns obtained contained diffused halos corresponding to the Laves phase reflections — again confirming the fine size and phase identification.

## DISCUSSION

The mechanical property data, yield strength, fracture energy (CVN), fracture toughness and hardness show that the formation of reverted austenite in Custom 450 results in a slight decrease in hardness and strength and with concomitant increase in fracture energy and fracture toughness. The austenite reversion apparently initiates at ~ $900°F$ (lowest temperature studied), and after a 4 hour age, 1% austenite is detected. The reversion reaction is sluggish at $900°F$. After 64 hour aging, only 3 to 4% austenite has formed. The reversion rate increases with time and temperature.

The increase in reverted austenite of 1 to 2% after a 4 hour age at $900°F$ and $1000°F$ has resulted in almost 100% increase in toughness with a relatively small decrease in yield strength, 25%. In the range studied, $900°F$ — 4 hours to $1050°F$ — 4 hours, the reverted austenite increased about 15% with an increase of ~ 200% in fracture energy but only a ~ 20% decrease in yield strength. The fine structure study of these specimens showed the size of the age hardening precipitates to be extremely fine, indicating that the small decrease in strength accompanied by a considerable increase in toughness is not due to precipitate growth and coalescence but due to austenite reversion.

The reverted austenite formation occurs along the martensite lathes in a ribbon-like morphology and improves the toughness. Pampillo and Paxton[5] have recently reported on a similar study on effect of reverted austenite on the mechanical properties and toughness of 12 Ni and 18 Ni maraging steels. They have shown that the reverted austenite in these maraging steels had no detrimental effect on properties and even improved the properties when precipitated along martensite lathe boundaries.

## SUMMARY

Reverted austenite has a significant effect on the mechanical properties and toughness of Custom 450. Yield strength and hardness decrease, and fracture energy and fracture toughness increases with increased austenite reversion. The improved

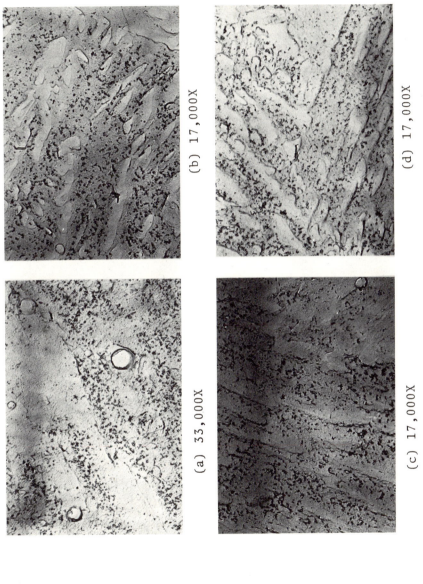

(a) 33,000X

(b) 17,000X

(c) 17,000X

(d) 17,000X

Fig. 6.  Carbon extraction replica electron micrographs of Custom 450 showing precipitate size.  Solution treated 1900°F, ½ h, WQ and aged as shown below: (a) 900°F – 64 h, AC  (b) 1000°F – 64 h, AC  (c) 1050°F – 64 h, AC  (d) 1000°F – 200 h, AC  (Reduced 20% for reproduction.)

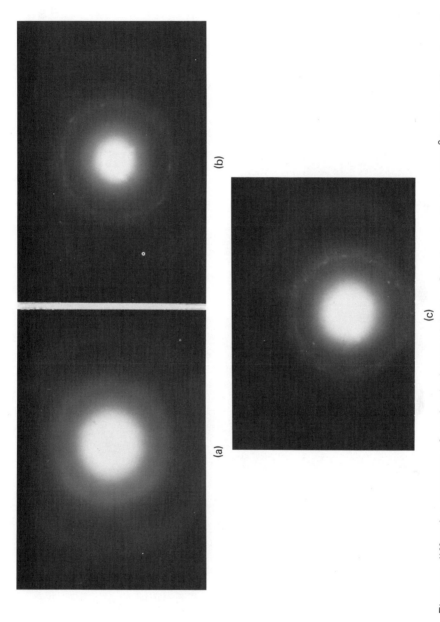

Fig. 7. Electron diffraction patterns of age-hardening precipitate. Solution treated at 1900°F, ½ h, WQ and aged as shown: (a) 900°F, 64 h, AC (b) 1000°F, 64 h, AC (c) 1050°F, 64 h, AC. Note the change from "halo" type pattern in (a) to "spotty" patterns in (b) and (c). The patterns for the precipitate aged at 900°F, 4 h, AC is similar to (a).

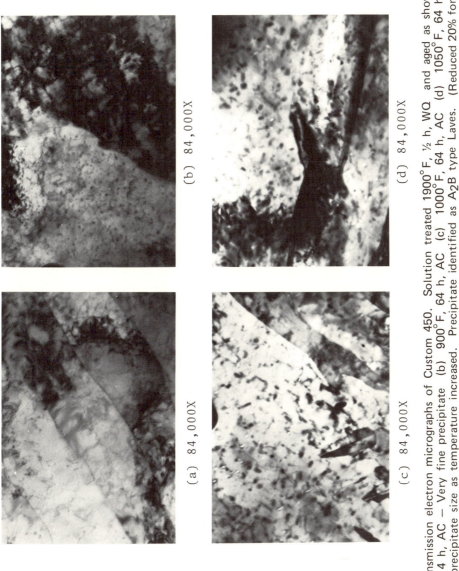

(a) 84,000X          (b) 84,000X

(c) 84,000X          (d) 84,000X

Fig. 8. Transmission electron micrographs of Custom 450. Solution treated 1900°F, ½ h, WQ and aged as shown below: (a) 900°F, 4 h, AC — Very fine precipitate (b) 900°F, 64 h, AC (c) 1000°F, 64 h, AC (d) 1050°F, 64 h, AC' Note increase in precipitate size as temperature increased. Precipitate identified as $A_2B$ type Laves. (Reduced 20% for reproduction.)

toughness is attributed to the ribbon-like austenite formed along the martensite lathe boundaries during reversion. Reversion initiates in the vicinity of 900°F, evidenced by 1% austenite after 900°F, 4 hour age. The rate of reversion increases with increasing temperature. The age-hardening phase has been conclusively identified as Laves, $M_2N_b$ type. The Laves precipitate in 900°F, 4 hour aged specimen is estimated to be ~ 20 A. The Laves phase coarsening rate with temperature and time, studied in this work, is negligible.

## FURTHER COMMENTS

This study is an excellent example of the usefulness of the XRD technique to measure austenite, retained or reverted, as a tool for quality control and research problems. It is easily seen that this technique can be applied to a multitude of alloys, such as:

(1)    Tool steels — the effectiveness of the austenitizing and tempering cycles in transforming retained austenite to imtempered austenite. The retained austenite generally cannot be detected from hardness and from metallographic studies.

(2)    Maraging Alloys — the age hardening response in most of these alloys as similar to that described for Custom 450.

(3)    Austenitic Stainless Steels — the austenitic pseudo martensite transformation via cold working can be easily followed.

(4)    Magnetic Alloys — Fe-Ni Alloys heat treatment and working can cause austenite to martensite transformations. The processing can be characterized to prevent occurrence.

(5)    The method cannot distinguish between martensite and ferrite, nor pseudo martensite and delta ferrite in austenitics. These features are easily distinguishable via metallographic techniques.

The XRD technique is a powerful tool but should be regarded as only one of the many available to a metallurgist.

It is important for the metallurgist to have a clear understanding of an alloy's physical metallurgical behavior, and the inter-related relationships of microstructural features to mechanical properties to effectively evaluate a particular behavioral characteristic.

## REFERENCES

1.    U.S. Patent # 3,574,601, "Corrosion Resistant Alloy", (April 13, 1971).
2.    *Custom 450 Data Booklet,* CarTech (1971).
3.    *Simplifying Stainless Steel Selection with Carpenter's Selectaloy Method,* Carpenter Technology Corporation (1969).
4.    *Manual on Electron Metallography Techniques,* ASTM STP 547, (1973).
5.    C.A. Pampillo and H.W. Paxton, "The Effect of Reverted Austenite on the Mechanical Properties and Toughness of 21 Ni and 18 Ni (200) Maraging Steels". *Met. Trans.,* Vol. 3, p. 2895 (1972).

## APPENDIX I

## PROCEDURE FOR X–RAY DIFFRACTION
## MEASUREMENT OF RETAINED AUSTENITE

The following procedure is used at Carpenter Technology Corporation's Research Laboratory to measure the amount of retained austenite by an X-ray diffraction technique using the G.E. XRD–5 diffractometer[1] :

(1)  Radiation — generally V filtered CrK$\alpha$ radiation is employed. However, Zr filtered MoK$\alpha$, Fe filtered CoK$\alpha$, or CuK$\alpha$ with a diffracted beam monochromator can be used.

(2)  X–Ray Optics — The size of the primary beam slit is chosen to have the sample surface area greater than the area irradiated by the primary beam.
     A medium resolution soller slit is employed. The receiving slit is removed so the counter tube will receive all the diffracted intensity. The wide opening (see below) increases the background but the total peak intensity is also increased significantly. The K$\beta$ filter or the diffracted beam monochromator is inserted in front of the counter tube.

(3)  A rapid scan, $2°$ ($2\theta$) per min., can be made to determine phases present in sample and to determine whether the martensitic and/or austenitic phase exhibits preferred orientation.

(4)  Using CrK$\alpha$ radiation, the (110) M (200) M, (211) M, (111) A, (200) A, (220) A peak profiles are obtained.
     The diffractometer's operation is semi-automated and can be programmed to slow scan or step scan through the angular range containing the austenite/martensite reflections. The diffractometer slews at $72°$ ($2\theta$) per minute to the next angular range. In the slow scanning mode ( $.25°$ or $.5°$ ($2\theta$) per minute) chart recordings are made of the peak profiles. The integrated intensity of each peak is then measured using a planimeter. In the step scanning mode, the step size and counting rate are chosen to obtain a well defined peak profile in a reasonable amount of time. The count rate at each angular step is automatically outputted to a teletype and punched on a paper tape. The integrated intensity, corrected for background, is obtained via a computer program. This method along with an automatic sample changer allows the diffractometer to operate unattended, especially from 5 p.m. to 8 a.m. If a small computer was added the intensity data could be entered directly, and the percent austenite in each sample would be the simplest output format.

(5)  The I.I. for each peak is corrected by the appropriate "R" factor, (described below),

$$I.I._{(hkl)} / R_{(hkl)} = N_{(hkl)}$$

(6)  The N values for each phase are averaged, $\bar{N}_M$ and $\bar{N}_A$, respectively. The percent austenite is calculated from $\bar{N}_A / (\bar{N}_M + \bar{N}_A)$ * 100.

(7)  The percent austenite is calculated to the nearest tenth of percent, but reported to the nearest whole percent. The precision of the measurement is ± 3% of the amount present.

(8)  The technique of using the three reflections from each phase, and averaging the calculated N values has been found to minimize the effects of preferred orientation.

The above procedure is based on the integrated intensity technique developed by Averbach and Cohen[2] and applied to counter-diffractometers by numerous investigators.[2–7]

The calculations are made using the formula for the integrated intensity (I.I.) of a particular phase;

$$\text{I. I.}_{(hkl)} = K * /FF/ * (\text{L.P.})\ e^{-2M} * A\ (\theta)\ *m * \frac{V}{v^2}$$

where:

| | | |
|---|---|---|
| I. I. $^{hkl}$ | = | Integrated intensity |
| K | = | Constant for given experiment |
| /FF/ | = | Structure factor times it complex conjugate |
| L.P. | = | Lorentz-polarization factor |
| m | = | Multiplicity factor |
| $e^{-2M}$ | = | Debye - Waller Temperature factor |
| A $(\theta)$ | = | Absorption factor (constant value for angles of diffractometer). |
| V | = | Volume fraction of phase |
| v | = | Volume of phase unit cell |

Since K and V are the only unknown terms, the rest can be equated to "R" and the above equation rewritten as:  I.I. (hkl) = K * V * $R_c$ (hkl) = K * $V_A$ * $R_A$ (hkl);  assuming $V_M + V_A = 1$, and K = K, then:

$$\frac{\text{I.I.}_M\ (hkl)}{R_M\ (hkl)} + \frac{\text{I.I.}_A\ (hkl)}{R_A\ (hkl)} = 1$$

I.I. (hkl) can be measured from strip chart recording or from computer integration of the peak profiles, i.e., scaler output connected to computer or to a teletype/punch tape.  The R(hkl) are calculated using a computer program with corrections for alloying and lattice parameters determined from $2\theta$ positions.

Factors and assumptions made in the calculation of the R factor are as follows:

/FF/          = $4f^2$ for martensite and $16f^2$ for austenite where f = atomic scattering factor * weight fraction of the alloying element — anamalous scattering factor.

(L.P.)        = $\dfrac{1 + \cos^2\theta}{\sin^2\theta \cos\theta}$, the Bragg angle for each reflection, is calculated from actual $2\theta$ position.

$e^{-2M}$     = $e^{-B}\ \dfrac{\sin\theta}{\lambda}$ is calculated in the computer program

The $a_o$ of martensite and the $a_o$ of austenite vary with alloying content.  The $2\theta$ positions of the reflections are determined and the $a_o$ for each phase calculated in the computer program.

Further assumptions must be made:
(1)   The martensite and austenite phases are randomly oriented.
(2)   The martensite and austenite phases are randomly dispersed throughout the sample.

Preparation of the sample surface can be critical. An acceptable sample surface can be prepared by metallographically polishing and etching. The sample surface should be representative of the microstructure.

## REFERENCES

1. Averbach, B.L., and Cohen, M., *Trans. A.I.M.M.E.,* Vol. 176, pp. 401–415, (1948).
2. Miller, R.L., *ASM Trans. Quart.,* 57, December, pp. 892–899 (1964).
3. Oglive, R.E., *Norelco Reporter,* Vol. VI, Number 3, p. 60 (1959).
4. Littman, W.E., MS Thesis, M.I.T. (1952).
5. Martin, J.A. Private Communication.
6. Erad, H.R., *Advances in X-ray Analysis,* Vol. 7, p. 256 (1963).
7. Durnin, J., and Ridal, K.A., *J.I.S.I.,* p. 60 (Jan. 1968).
8. Cullity, B.D., "Elements of X-ray Diffraction", (1956).

## APPENDIX II

## XRD — RETAINED AUSTENITE

Accuracy and Precision

I. Random sample — SRM 483 Certified 3.4%

| | |
|---|---|
| Average of 10 measurements | 3.3% |
| Standard deviation | ± 0.1% |

Accuracy for random $\sim$ 3% of amount present
Precision for random $\sim$ 3% of amount present

II. Heavily oriented sample

% Austenite via metallographic point count                38%

| XRD Measurements | % Austenite Averaging R | % Austenite by Pair Method |
|---|---|---|
| Transverse Section | 51% | 47% |
| Longitudinal Section | 30% | 42% |
| 45° Section | 41% | 40% |
| Average Value | 40.6% | 43% |
| Standard Deviation ± | ± 10.5% | 3.6% |
| Composite Sample Rotated and Tilted | 42% | |

Sphere — ground and polished              39%
Accuracy can be stated as ±              26%
Example: if 5% $\gamma$ measured, the value could be ranged between 4 to 6%

# METALLOGRAPHIC CONTROL OF HEAT TREATMENT

E.G. Nisbett*

## INTRODUCTION

The aspect of metallography as a quality control tool in heat treatment will be discussed principally from the standpoint of its use in controlling the heat treatment of large open die steel forgings intended for use in a broad spectrum of industries. Some of the controls involve well established techniques such as McQuaid-Ehn grain size estimation and are employed as a routine production check; others are used in an "after the event" capacity to seek the causes and point to the correction of specific heat treatment problems. A third use involves heat treatment development. Inevitably a part of this activity infringes on the field of failure analysis, especially as this applies to work in process or heat treatment equipment.

## GRAIN SIZE

Grain size determination is probably the most widely practiced metallographic control of heat treatment because of the effects this microstructural feature can impart (Table 1). The procedures for carrying out this test are described in some detail in ASTM E112[2] "Standard Methods for Estimating the Average Grain Size of Metals," and attention is drawn to Annex A3 of this specification.

When the austenitic grain size of a steel is sought, one of the outlining procedures described in ASTM E112 can be used, but it is not uncommon for a purchase order to reference the McQuaid-Ehn test specifically. While this does work well in many instances (Fig. 1), it is not without problems with some of the higher alloyed materials, such as the example of a large 9% chromium-1% molybdenum alloy steel forging to ASTM-A182, grade F9 shown in Fig. 2. McQuaid-Ehn and controlled cooling outlining test results, together with the normalized and tempered structure, are shown in Fig. 3.

The final, or heat-treated grain size, in addition to being simpler to obtain, often gives the user of the forging a better idea of its capabilities and increasingly it is this parameter which is looked for. It is sometimes helpful in terms of perspective to consider the part which is represented; for example, the large turbine bucket wheel shown in Fig. 4 when one examines the metallographic sample. The alloy in this case was of the nickel-chromium-molybdenum-vanadium type with the following composition

| C | Mn | S | P | Si | Ni | Cr | Mo | V |
|------|------|------|------|------|------|------|------|------|
| 0.29 | 0.30 | 0.0 | 0.0 | 0.07 | 2.75 | 1.50 | 0.55 | 0.09 |

in the water quenched and tempered condition.

*National Forge Company, Irvine, Pennsylvania, USA.

## TABLE 1

### TRENDS IN HEAT-TREATED PRODUCTS

| PROPERTY | COARSE-GRAIN AUSTENITE | FINE-GRAIN AUSTENITE |
|---|---|---|
| Hardenability | Deeper Hardening | Shallower Hardening |
| Toughness | Less Tough | Tougher |
| Distortion | More Distortion | Less Distortion |
| Quench Cracking | More Prevalent | Less Prevalent |
| Internal Stress | Higher | Lower |

### FOR ANNEALED OR NORMALIZED PRODUCTS

| | | |
|---|---|---|
| Machinability (Rough) | Better | Inferior |
| Machinability (Fine Finish) | Inferior | Better |

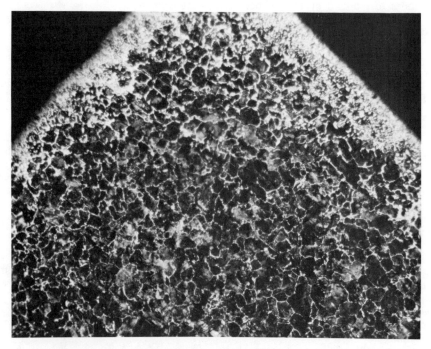

Fig. 1.  Typical McQuaid-Ehn grain size sample in AISI 4140 material.  100X Etchant;  2% Nital.

Fig. 2. Large mandrel forging in A182 grade F9 cooling prior to post forge heat treatment.

The grain size was estimated to be 7–9 in the ASTM grain size range. The microstructure is shown, after etching the sample in 2% Nital (Fig. 5a) and Villella's reagent (Fig. 5b).

Grain size in austenitic stainless steels is sometimes requested to be held within specific limits, but in view of the difficulty of controlling this, particularly in the larger forgings, there is some reluctance to comply on the part of the forgemaster. An example shown in Fig. 6 was from a cold worked periscope tube in type 304 austenitic steel. The sample was etched electrolytically in 10% oxalic acid. The effects of surface cold working during sample preparation help in assessing the austenite grain size.

## SERVICE RELATED CRITERIA

A further aspect of production heat treatment control concerns the completeness of transformation in low alloy steels. A typical example of this type of control is shown in Fig. 7 for a water-quenched and tempered nickel-molybdenum vanadium alloy steel generator shaft, with the following analysis:

| C | Mn | S | P | Si | Ni | Mo | V |
|------|-----|-----|-----|-----|------|------|------|
| 0.24 | 0.4 | 0.0 | 0.0 | .22 | 3.18 | 0.35 | 0.08 |

This type of control also gives the opportunity to check on possible temperature excursions over the lower critical temperature during the temper cycle.

As a result of concerns about stress corrosion cracking of austenitic steels, such as types 304 and 316 intended for nuclear power plant use, the U.S. Atomic Energy

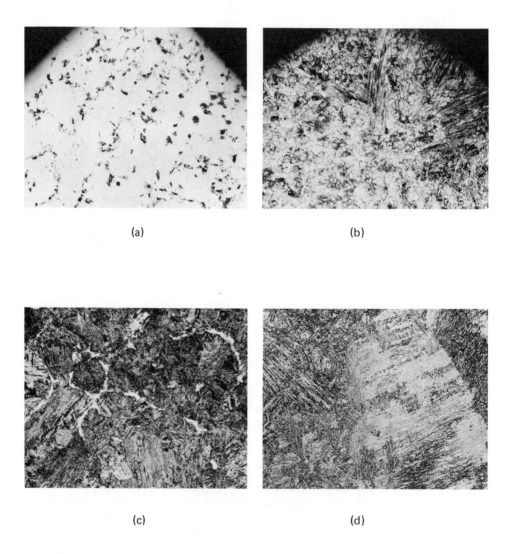

(a)                                              (b)

(c)                                              (d)

Fig. 3.A. McQuaid Ehn grain size sample in A182 grade F9. Note unsatisfactory grain size definition. 100X; Etchant 2% Nital. B. Same sample as in Fig. 3A. Grain size resolution not properly defined. 100X; Etchant Villella's reagent. C. Attempted grain outlining by step cooling in A182 grade F9 material. 100X; Etchant Villella's reagent. D. Heat-treated microstructure in A182 grade F9 after normalize & temper heat treatment, showing coarse grain size. 100X; Etchant Villella's reagent. (Reduced 45% for reproduction.)

Fig. 4.  Large bucket wheel forging being conveyed to quench tank.  Mechanical test prolongations are located at the inner hub and outer rim.

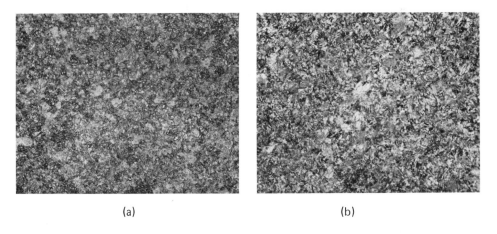

(a)                                           (b)

Fig. 5.A. Microstructure of water quenched and tempered bucket wheel forging. Heat treated grain size ASTM 6–8. 100X; Etchant 2% Nital.  B.  Sample from the same forging as Fig. 5A using alternate etchant for grain size determination.  100X; Etchant Villella's reagent.  (Reduced 45% for reproduction.)

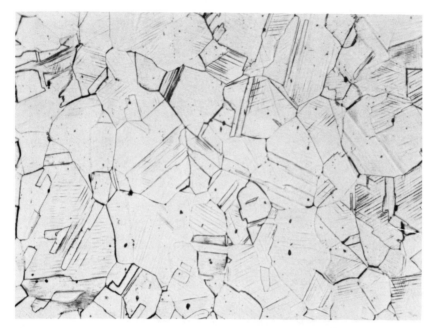

Fig. 6. Grain size, ASTM 2 from a forging in type 304 austenitic stainless steel, cold worked after solution treatment. 100X; Etchant 10% oxalic, electrolytic.

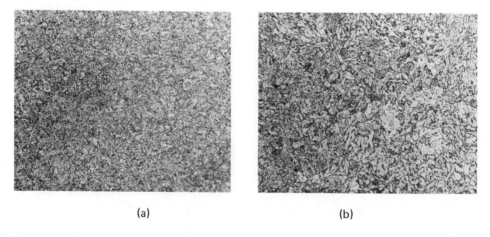

(a)                                                                    (b)

Fig. 7.A. Microstructure of a sample from a water quenched & tempered generator shaft showing complete transformation. 100X; Etchant 2% Nital. B. Sample 7c at higher magnification confirming absence of retained austenite. (Reduced 45% for reproduction.)

Commission regulatory guide 1.44[3] was issued on the control of sensitized stainless steel. This guide proposed the use of Method E-copper-copper sulphate-sulphuric acid test, or Method A-oxalic acid etch test as described in ASTM-A262[4] (Recommended Practices for Detecting Susceptibility to Intergranular Attack in Stainless Steels) as methods to verify that the material would be non-susceptible to intergranular stress corrosion. The original intent was to apply this screening to material which had not been water quenched after solution heat treatment, but in many cases it has been applied across the board. Ditch structures are unacceptable when completely surrounding an individual austenite grain. Figure 8 shows a marginal dual structure while Fig. 9 demonstrates the beneficial effect of a reheat treatment, by taking grain boundary carbides into solution.

Surface decarburization is another area where metallographic control of heat treatment is required by specifications such as MIL-S-5000[5] and AMS 6429[6]. This effect can, of course, be deleterious in finished parts by reducing fatigue strength and surface hardness. By the same token, surface carburization can occur inadvertently in some controlled heat treatment atmospheres, and this too may have adverse effects. The following example was found in the course of failure analysis, but it does reflect the assistance that metallography can offer in heat treatment control. High tensile strength bolts (Fig. 10) were the subject. These bolts appeared to be extremely brittle, with a tendency to fail under the head (Fig. 11). The bolt surface was slightly carburized as shown in Fig. 12, a condition coming from the heat treatment furnace atmosphere. A hardness traverse (Fig. 13) through the carburized skin helps explain the brittle behavior of the bolts.

Fig. 8. Austenitic type 304 stainless steel sample from a large forging showing marginal dual (step-ditch) structure at the grain boundaries when etched per ASTM A266 Method A. 100X; Etchant 10% oxalic electrolytic.

Fig. 9. Sample from the same forging as Fig. 8 after solution heat treatment showing step effect at the grain boundaries.   100X;   Etchant 10% oxalic, electrolytic.

Fig. 10.   Bolt similar to the type which failed during assembly.

Fig. 11. Shank of bolt which fractured at the underside of the head.

Fig. 12. Section through quenched & tempered AISI 4140 bolt shank showing surface carburization leading to a brittle skin. 100X; Etchant 2% Nital.

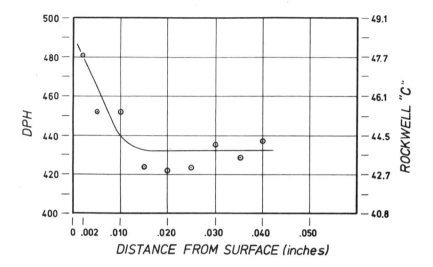

Fig. 13.   Microhardness survey through surface skin of fractured bolt.

## SURFACE HARDENING

Improved service performance as a result of surface hardening ensures a steady use of this class of industrial processing and many of the techniques utilized metallographic examination in initially developing the processes, if not in the day-to-day operation.

In the case of induction or flame hardening the metallographic techniques come into play in setting up and adjusting the equipment, particularly in obtaining the required transition zone characteristics.   Hardness testing then looks after the production control aspect although surface etching may be useful in some circumstances. Macro and micro examination of course again is used in the course of resolving problems.

The example shown in Fig. 14 shows a typical setup for a repetitious induction hardening operation for crankshafts in AISI 5046 material.   The typical hardened zone pattern from a bearing section is shown in Fig. 15.   The structures of the originally quenched and tempered core structure (Fig. 16a), the transition zone for the induction hardened case (Fig. 16b), and the fully hardened case itself (Fig. 16c) reflect the changes in hardness pattern seen in the hardness curve (Fig. 17).   Heating conditions were controlled to avoid an abrupt hardness drop off, by introducing a relatively wide intercritically quenched transition zone.

Nitriding as a surface enrichment process is mentioned here to represent other such processes as carburizing and carbonitriding, etc.   Although hardness testing is again used as an acceptance and quality assurance tool, in many cases small samples of the same material are included with the furnace load for control purposes, so that the depth of case can be measured.   Since there is no sharp cutoff in these

Fig. 14.  Induction hardening setup for diesel crankshafts.

Fig. 15.  Macro section through induction hardened crankshaft pin, showing depth and extent of hardened surface.

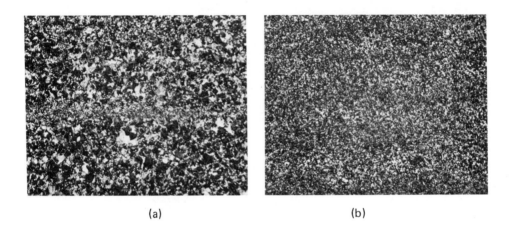

(a)                                                    (b)

(c)

Fig. 16A. Case structure of quenched and tempered AISI 5046 crankshaft forging.
B. Intercritically heat-treated transition zone of induction hardened zone. C.
Lightly tempered martensitic surface zone of induction hardened crankshaft. All at
100X and etched in 2% Nital. (Reduced 45% for reproduction.)

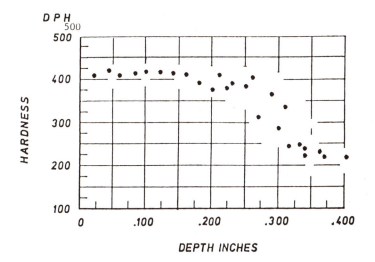

Fig. 17. Hardness survey through an induction hardened and subsequently tempered crankshaft bearing in AISI 5046 material.

diffusion-type processes, some agreed upon and necessarily arbitrary cutoff point has been used in determining case depth.

A large nitriding furnace is shown in Fig. 18, with a crankshaft charge being made up. Areas which do not require hardening are coated with a special paint to mask the nitriding effect. The control samples (Fig. 19) in AISI 4130 material, normalized and tempered, are sectioned after nitriding and prepared for hardness traverse and case depth checks (Fig. 20). Since the nitrided ferrite has a higher hardness (about 55 $R_C$) than the nitrided pearlite (about 45 $R_C$), the microhardness survey (Fig. 21) shows much scatter. Aluminum-containing steels, such as Nitralloy 135, respond well to the nitriding process to give high surface hardnesses of over 67 $R_C$ (Fig. 22). The material, quenched and tempered before nitriding, gives an ultimate tensile strength in the case of about 125 KSI (860 MPA), with a tempering temperature of 1200° F. (650° C.). There is little, if any, change following the nitriding cycle. The hardness test traverse results through the case of a nitralloy sample are reproduced in Fig. 23.

Another production heat treatment which relies on metallography for control is spheroidization. Forgings are often supplied in a spheroidized condition, for subsequent heat treatment elsewhere, as for example, steel rolls made from AISI-E52100 material. Figure 24 shows the spheroidized microstructure of a typical roll product in this grade.

PROBLEM SOLVING

Response to heat treatment sometimes strays away from the intended results and one must look to the variables to explain the problem and seek a solution. While mechanical test results, hardness testing, chemical analysis, machining difficulties

Fig. 18. Large nitriding furnace being charged with crankshafts. Grey areas have been painted to stop off nitriding effects.

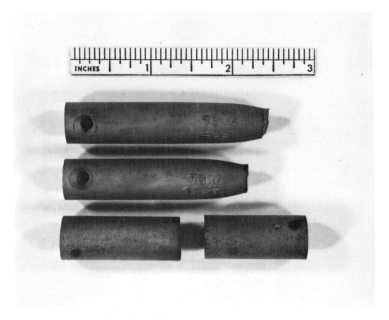

Fig. 19. Nitriding control samples.

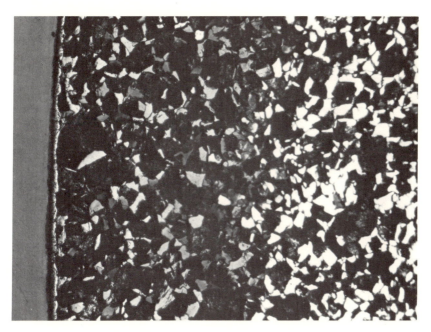

Fig. 20.  Section through nitrided sample on normalized and tempered AISI 4130 type forged crankshaft material.  Nitride cycle was 72 hours with maximum temperature of 1000° F.  (540° C.).

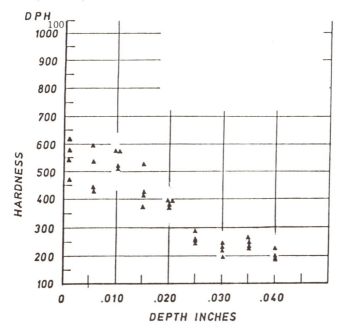

Fig. 21.  Nitride case depth of AISI 4130.  Normalized and tempered, nitrided 172 hours at 1000° F. (540° C.) max.

Fig. 22. Typical nitrided case in quenched and tempered Nitralloy type 135 material. 100X; Etchant 2% Nital.

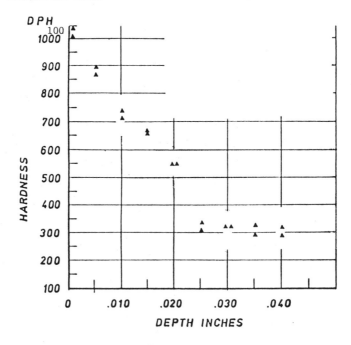

Fig. 23. Nitride case depth in Nitralloy 135 modified. Quenched and tempered, nitrided 72 hrs. at 1000°F (540° C.) max.

Figl 24. Typical spheroidized structure of forged AISI 52100 material.  100X;
Etchant 2% Nital.

or non-destructive examination may indicate that there is a problem, such results by
themselves may not show how it can be corrected.  Metallography can often fill in
the gaps so that an informed reheat treatment or replacement can be made.  The
tension test results can indicate problems in quenching.  A lower yield strength than
would have been expected could indicate a slack quench.  Such a situation is illustrated in Fig. 25, which shows the microstructure of a water quenched and tempered Ni-Cr-Mo-V steel after the original heat treatment.  The structure after reheat
treatment, including double tempering, is shown in Fig. 26, and the improved structure reflected better mechanical properties.

Poor ductility in the tension test may be caused by adverse orientation of the
test sample relative to the major working direction of the forging, excessive non-
metallic inclusions and, less frequently, by severe overheating during the forging cycle.

Special etching techniques, such as the nitro-sulphuric etch used in the following series, can help pinpoint the latter problem.  Figure 27 shows the abnormal
tensile specimen fracture surface and the microstructure of the quenched and temi
pered nickel-chromium-molybdenum-vanadium alloy steel is illustrated, after Nital
etching, in Fig. 28, while the grain boundary damage is seen in Fig. 29, after using
the nitro-sulphuric etchant.

Coarse grain size in austenitic stainless steel forgings is known to adversely affect ultrasonic examination.  An example of this is the large 2% Ni-Cr-Mo-V steel
crusher shaft shown in Fig. 30.  The forging had received an unsuitable post forging
heat treatment, and it was not possible to conduct a meaningful ultrasonic examination.  The microstructure causing this problem is shown in Fig. 31.  Reheat treatment
of the shaft refined the structure (Fig. 32) and effected a satisfactory cure.

Fig. 25. Mixed microstructure of Ni-Cr-Mo-V low alloy steel showing the effects of slack quenching. 100X; Etchant 2% Nital.

Fig. 26. Uniform tempered martensite structure of forging in Fig. 25 after reheat treatment. 100X; Etchant 2% Nital.

Fig. 27.  Abnormal brittle tension test sample in Ni-Cr-Mo-V alloy steel, quenched and tempered.

Fig. 28.  Coarse grained microstructure from tension test sample.  100X;  Etchant 2% Nital.

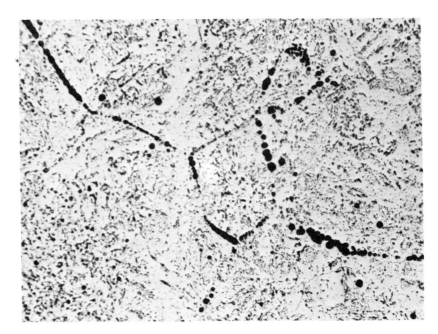

Fig. 29. Micrograph from tension test sample showing grain boundary damage caused by overheating during forging. 100X; Etchant nitro-sulphuric acid.

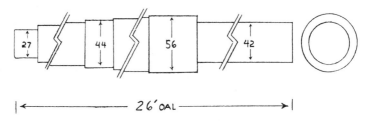

Fig. 30. Sketch of crusher shaft forging.

Austenitic steel forgings, however, are not so easily dealt with, as in one, rather frustrating, not to mention expensive, example shown in Fig. 33. The part was a special tubular fitting for nuclear service and was made as a solid forging in type 304 stainless steel. The forging was bored before solution heat treatment, and an ultrasonic examination involving straight beam and shear wave testing was required. The forging was found to be acceptable except for a local area in the tubular section which yielded strong shear wave responses exceeding the calibration level. *In situ* metallography (Fig. 34) suggested that a local patch of coarse grain size was responsible and a portion of the surface was milled away to permit limited "through the section" metallography (Fig. 35). Nevertheless, acceptance of the part was refused. On cutting up the forging, no condition other than the coarse grain size was found in the area of the ultrasonic indications. This is illustrated in Fig. 36. Tension and bend tests from the coarser grained material showed no ill effects. The coarse grained area was probably formed in an area lightly cold or warm worked prior to solution treatment.

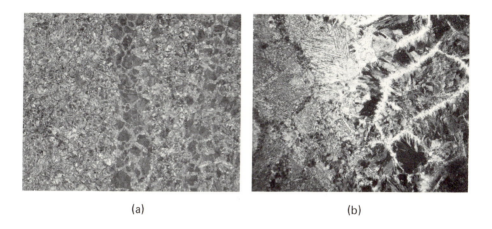

(a)                                        (b)

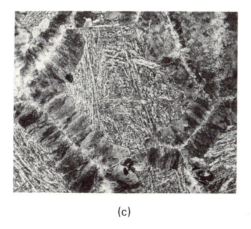

(c)

Fig. 31    Coarse irregular structure in low alloy crusher shaft forging caused by improper post forge heat treatment. 10X; Etchant 2% Nital. B. Detail of complex
mixed microstructure of crusher shaft sample. C. Further detail of shaft microstructure. Variation caused by alloy segregation and heat treatment, and resulting structure interfered with ultrasonic examination. 100X; Etchant 2% Nital. (Reduced 45%
for reproduction.)

Fig. 32. Uniform microstructure of the shaft after correct normalize and temper post forge heat treatment, giving acceptable structure for ultrasonic examination. 100X; Etchant 2% Nital.

Fig. 33. Austenitic containment penetration forging.

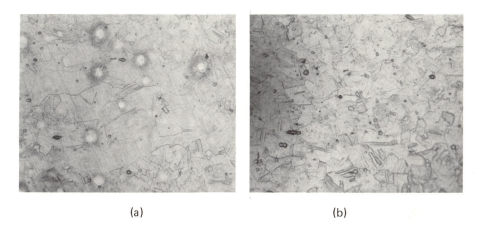

(a)                                                    (b)

Fig. 34 .A. Coarse grained structure from local area of the surface of a penetration forging.  B.  Average grain size of penetration forging from machined surface.  Both at 100X and etched in 10% Oxalic Acid,  Electrolytic. (Reduced 45% for reproduction.)

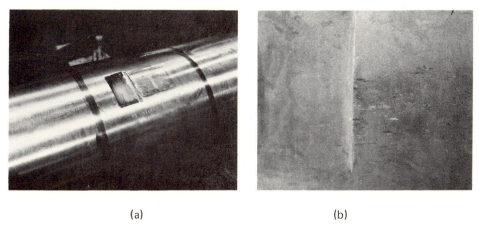

(a)                                                    (b)

Fig. 35. A. Surface of penetration showing area locally milled away to investigate local area of large grain size.  B.  Etched surface at milled flat on outside diameter of penetration showing elongated local large grains. (Reduced 45% for reproduction.)

(a)                                                          (b)

Fig. 36.A. Section through wall of austenitic penetration forging showing local coarse grain extending from outer surface. B. Section through wall of forging at fringe of coarse grain area. (Reduced 45% for reproduction.)

Martensitic stainless steels, such as AISI 410, in heavier section sizes can sometimes give magnetic particle indications which are indicative of a metallographic condition rather than demonstrating the presence of a crack. Carbides outlining ferrite stringers cause strong magnetic particle indications during magnetic particle examination. The effect can be minimized by solution treating the material at temperatures of about 2000° F. (1095° C.), followed by oil quenching. Typical microstructures showing the delta ferrite outlined by carbide are presented in Fig. 37.

Although metallographic structures can be developed for optimum machinability, such as the spheroidized condition just described, resolving apparent machining difficulties for a given structure or heat treatment condition can be very difficult, particularly since other variables may not be well defined. It is always prudent to take note of feedback from the machine shop, as in this example. Local machining abnormalities lead to an investigation of a low carbon steel forging, and this effect (Fig. 38) caused by a localized patch of steel to a much higher carbon content was found. The microstructure (Fig. 39) shows the difference in the base microstructure and that of the foreign material. The foreign material apparently dropped into the molten steel during the teeming operation.

## HEAT TREATMENT DEVELOPMENT

Since, as a general statement, the heat treatment of steels is phase change related, metallography can be expected to assume a major role in heat treatment development. Need is the main spur in any development program, and heat treatment development is often done quite hurriedly to solve a particular problem, as in the last example.

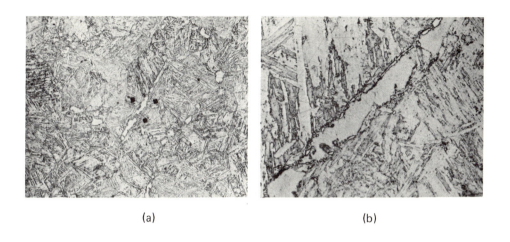

(a)                                      (b)

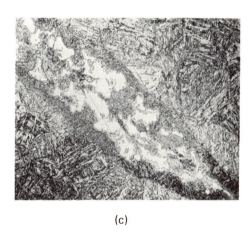

(c)

Fig. 37 .A. Sample from quenched and tempered AISI 410 stainless steel forging showing local delta ferrite stringer. Microhardness impressions were for location purposes. 100X. B. Delta ferrite stringer outlined by carbides which tend to cause magnetic particle examination indications. 800X. C. Irregular shaped delta ferrite phase in quenched & tempered AISI 410 forging. Carbide outline also causes broad magnetic particle indications. 100X. All etched in 2% Nital. (Reduced 45% for reproduction.)

Fig. 38.  Macro sample from carbon steel forging showing dark etching zone of foreign material.

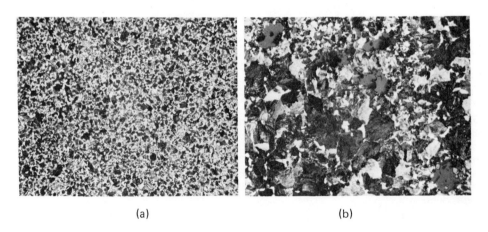

(a)                                              (b)

Fig. 39.A. Typical ferrite and pearlite structure in base forging shown in Fig. 38.
B.  Coarser grained higher carbon foreign material in dark etching zone of Fig. 38.
Both at 100X;  etched in 2% Nital. (Reduced 45% for reproduction.)

The intercritical heat treatment of carbon and certain low alloy steels is an example of a larger term program which has been reported elsewhere[7,8,9]. In this case metallographic examination of test samples was essential in determining the effects of heat treatment cycle variations on the microstructure and how shifts in the mechanical properties were related to microstructure changes. The program was started in order to explore the heat treatment possibilities for the development of acceptable impact properties in carbon and certain pressure vessel related low alloy steels, within the constraints of existing chemical analysis limits.

Generally speaking, the carbon steel pressure vessel forging specifications had for many years restricted carbon content to 0.35% maximum, with a manganese maximum of 0.90%. For low temperature applications, a modified chemistry with carbon content restricted to 0.30% maximum and manganese boosted to 1.35% maximum was used, but with the shift to the Charpy V notch test from the U notch specimen, even this material was found to have severe shortcomings in large section sizes. In such carbon steels, quenching and tempering has long been recognized as a method of enhancing Charpy impact properties, and from the section sizes involved, it was readily apparent that the improvement was associated with grain size, and perhaps pearlite refinement rather than the production of martensitic or bainitic structures.

It was from the observation that multiple quench cycles followed by tempering often lead to structure refinement and reduction in grain size that the idea of austenitizing between the upper and lower critical temperatures grew. The need in this case consisted of a batch of forgings which would not respond to heat treatment from temperatures above the $A_{C3}$. The difference in microstructure is evident from Fig. 40.

The intercritical heat treatment results in a significant reduction in grain size through the section of the heat-treated part, so that the improvement in impact results is not restricted to a relatively shallow skin, as tends to be the case with a conventional cycle austenitized above the upper critical. This is demonstrated by micrographs taken from a 22-inch (560 mm) diameter forged bar in carbon steel to ASTM A105, melted to fine grain practice. The original normalized and tempered microstructure at the center depth is shown in Fig. 41, with an ASTM grain size of 5–7. At the same location, after an intercritical double quench and temper cycle, the grain size was significantly reduced to ASTM 8–10, as is seen in Fig. 42.

The process also works well with an intercritical double normalize and temper technique. This is demonstrated in Fig. 43. These samples were also taken from the center locations of a 22" (560 mm) diameter forged bar, from a second heat of steel to ASTM A105. The microstructures of the intercritically double quenched and tempered material and the intercritically double normalized and tempered material appear to differ mainly in the distribution of the fine pearlite and carbide in the ferritic matrix.

A more dramatic picture of the refinement in grain size brought about by heat treatment is gained from Fig. 44, which compares the microstructure of a large piece forged to ASTM A350, grade LF2, and the same piece after an intercritical double quench and temper cycle.

CONCLUSION

This survey of how metallography can be used to control heat treatment of steel forgings has not touched on some of the more specialized areas such as tool

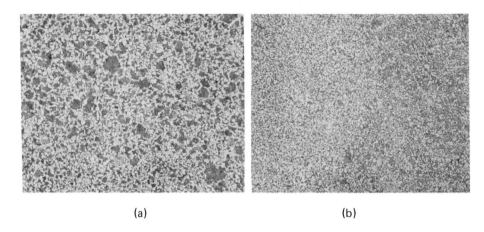

(a)                                                    (b)

Fig. 40.A. Quenched and tempered low carbon steel forging Fig. 40, showing grain size typically 5–7 ASTM. B. Intercritically double quenched & tempered forging – Fig. 40, reheat treated showing refined grain size ASTM 8–10. Both at 100X and etched in 2% Nital. (Reduced 45% for reproduction.)

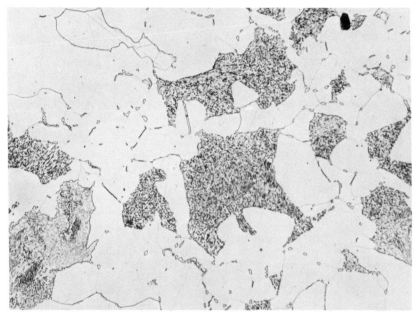

Fig. 41. Normalized & tempered low carbon-manganese steel forging to fine grain practice showing typical pearlite & ferrite structure at center of 22″ section. 400X. Etchant 2% Nital.

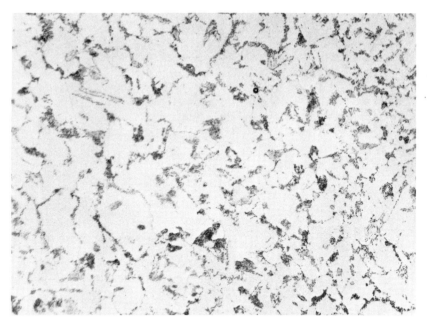

Fig. 42. Intercritically double quenched & tempered microstructure of forging shown in Fig. 44. Note pearlite distribution. 800X; Etchant 2% Nital.

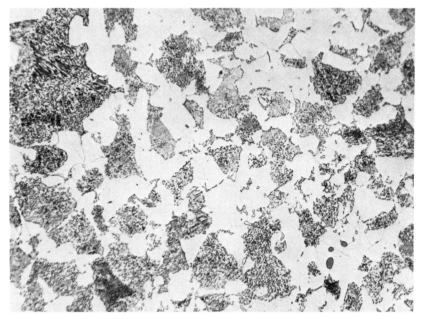

Fig. 43. Normalized & tempered microstructure of 22" diameter forged bar in low carbon manganese steel, at center location. 400X; Etchant 2% Nital.

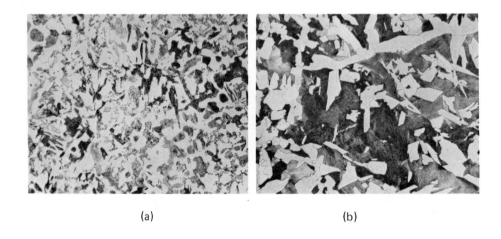

(a)                                              (b)

(c)

Fig. 44.A. Intercritically double normalized & tempered structure of forging in Fig.
44 also at center location. Note pearlite distribution as compared with Fig. 43.
(400X). B. Ferrite & Pearlite in microstructure of a large carbon steel forging, as
air cooled after forging. (100X). C. Structure of forging shown in Fig. 44B after
intercritical double quench & temper cycle. (100X). All etched in 2% Nital.
(Reduced 45% for reproduction.)

steel heat treatment, but it is hoped that the importance of metallography in maintaining and developing heat treatment quality has been adequately described.

## ACKNOWLEDGMENTS

The author wishes to thank National Forge Company for permission to write this paper as well as the many colleagues who assisted in its content and preparation.

## REFERENCES

1. *The Making, Shaping & Treating of Steels,* United States Steel Corporation, 8th Edition.
2. ASTM E112—74 "Estimating the Average Grain Size of Metals".
3. U.S. Atomic Energy Commission Regulatory Guide 1.44 "Control of the Use of Sensitized Stainless Steel" — (May 1973).
4. ASTM A262—75 "Recommended Practices for Detecting Susceptibility to Intergranular Attack in Stainless Steels".
5. MIL—S—5000 — Military Specification — "Steel, Chrome-Nickel-Molybdenum (E4340) Bars and Reforging Stock" Weapons Engineering Standardization Office, Naval Air Engineering Center, Philadelphia, Pennsylvania.
6. AMS 6429 "Steel Bars, Forgings & Tubing 0.80 Cr-1.8 Ni-.35 Mo-20 V (0.33—0.38C) Consumable Electrode or Vacuum Induction Melted", Aerospace Material Specification — Society of Automotive Engineers.
7. E.G. Nisbett, R.D. Asp and D.E. Morgan, "Improving the Notch Toughness of Nuclear Forgings in Carbon and Low Alloy Steels by Intercritical Heat Treatment", 8th International Forgemasters Meeting, Kyoto, Japan (1977).
8. E.G. Nisbett, "Factors Affecting the Toughness of Carbon and Low Alloy Steel Forgings for Pressure Vessel & Piping Applications", *Trans. A.S.M.E. Journal of Engineering, Materials & Technology,* 338—346 (October 1978).
9. E.G. Nisbett, R.D. Asp, D.E. Morgan, "Improving the Notch Toughness of Nuclear Forgings by Intercritical Heat Treatment", *Metals Progress*, Vol. 114, No. 2, 54—60 (July 1978).

# AUTHOR INDEX

## A

Aldinger, S., 151
Alford, N.A., 135
Alisanova, S., 19, 82
Allmand, T.R., 13, 52, 54, 60, 62, 78, 79, 81, 82, 84, 85, 86
Allsop, R.T., 54, 82
Andrew, R.C., 219
Anstead, R., 278
Aronovich, M.S., 54, 83
Asp, R.D., 327
Austin, G.W., 219
Averbach, B.L., 293, 295

## B

Baird, S.S., 278
Baker, T.J., 11, 12, 81
Bandi, W.R., 81
Bandyopadhyay, G.K., 55, 58, 84
Bardos, D.I., 221
Barrie, A., 135
Barteld, K., 82
Bartholme, W., 87
Bartz, G., 151
Baynes, A.D., 54, 55, 83
Bayer, J.L., 85
Bayre, W.W., 78
Belk, J.A., 85
Bellot, J., 12, 81
Benedicts, C., 19, 20, 78
Benko, D., 136
Benninghoven, A., 135
Beraha, E., 60, 87
Bergh, S., 53, 54, 56, 73, 80, 83, 84
Betteridge, W., 79
Beveridge, W.A., 77
Birkle, A.J., 84

Blaise, G., 135
Blank, J.R., 52, 79, 82, 85
Bloom, R.A., 80, 85
Bolsover, G.R., 19, 20, 81
Brandeis, H., 86
Brandemarte, A.V., 77
Bridge, J.E., Jr., 279
Brinen, J.S., 135
Brundle, C.R., 136
Brunet, J.C., 12, 81
Buhler, H.E., 62, 87, 151
Burke, K.E., 80
Burr, W.H., 78
Burstein, G.T., 135
Buzek, Z., 55, 84

## C

Cachard, A., 135
Campbell, W., 78
Carbonara, R.S., 135
Carney, D.J., 55, 84
Charles, J.A., 11, 12, 81
Chin, L.L., 84
Christian, A.F., 219
Clifford, D.A., 278
Cohen, M., 293, 295
Colby, B.N., 135
Cole, M., 85
Cole, N.C., 183
Coleman, D.S., 13, 62, 81, 85, 86
Comstock, G.F., 78
Coons, W.C., 278
Cottingham, D.M., 85
Crafts, W., 53, 54, 83
Cressman, R.N., 78
Cullity, B.D., 295

# SUBJECT INDEX

335